T5-AGX-369

GEOLOGICAL ESSAYS

This is a volume in the Arno Press collection

History of Geology

See last pages of this volume
for a complete list of titles

GEOLOGICAL ESSAYS

Richard Kirwan

ARNO PRESS

A New York Times Company

New York / 1978

Editorial Supervision: ANDREA HICKS

Reprint Edition 1978 by Arno Press Inc.

HISTORY OF GEOLOGY
ISBN for complete set: 0-405-10429-4
See last pages of this volume for titles.

Manufactured in the United States of America

Publisher's Note: This book has been reproduced from the best available copy.

Library of Congress Cataloging in Publication Data

Kirwan, Richard, 1733-1812.
Geological essays.

(History of geology)
Reprint of the 1799 ed. printed by T. Bensley for D. Bremner, London.
1. Geology--Early works to 1800. I. Title. II. Series.
QE25.K57 1977 550'.8 77-6523
ISBN 0-405-10444-8

GEOLOGICAL

ESSAYS.

By RICHARD KIRWAN, Esq.
F.R.SS. Lond. & Edin. M.R.I.A.
OF THE ACADEMIES OF STOCKHOLM, UPSAL, BERLIN,
MANCHESTER, PHILADELPHIA;
OF THE MINERALOGICAL SOCIETY OF JENA, &c.
AND INSPECTOR GENERAL OF HIS MAJESTY'S MINES IN THE
KINGDOM OF IRELAND.

LONDON:
PRINTED BY T. BENSLEY, BOLT COURT, FLEET STREET,
FOR D. BREMNER, (SUCCESSOR TO MR. ELMSLY) STRAND.

1799.

PREFACE.

GEOLOGY is the ſcience that treats of the various relations which the different conſtituent maſſes of the globe bear to each other. It at once unfolds and ſhews how to read the huge and myſterious volume of inanimate nature, of which MINERALOGY ſupplies the alphabet. The inſtruction it conveys is our ſureſt guide in reſearches for the various valuable ſubſtances buried under the earth's ſurface, and powerfully aſſiſts us in the ſublime inveſtigation of the hiſtory of the planet we inhabit. Singularly diverſified, and intricately complicated, as the local arrangement of ſubterraneous ſubſtances may appear, yet that its mazes are not without a clew, may readily be inferred, and with certainty concluded, from the practical ſkill which ſeveral miners are known to poſſeſs in many parts of the world. Uncombined, however, with any general theory, the knowledge thus attained, is generally imperfect, being circumscribed within the limits of the districts wherein their operations are exercised, linked with their peculiar circumstances, and more over, frequently darkened and perplexed with notions either falsely assumed, or erroneously generalized.

It is to men of far superior acquirements that Geology owes its origin and progress. John Gottlob Lehman first traced the genuine outlines of this science: Eminently skilled in general physics, practical mining, mineralogy, and chymistry, fully acquainted with the circumstances attending the situation of most minerals relatively to each other, in numerous and extensive tracts of different countries, he was enabled to deduce from multiplied observations some general conclusions, which have since, with few exceptions, been verified in all parts of the world.

This sagacious observer soon found that *priority* or *posteriority* of formation was a prominent feature distinguishing various elevations of the globe, and a circumstance demanding the strictest attention in determining the probable presence of particular minerals, as well as in devising and prescribing

ſcribing the mode of extracting them. Hence aroſe the denominations of PRIMARY or PRIMEVAL, and of SECONDARY mountains. And thus Geology was found connected with the ancient hiſtory of the globe.

The connexion thus diſcovered between the modern and ancient ſtate of the Earth ſoon excited the curioſity, and fired the imagination of ſome ſpeculative philoſophers, better verſed in mathematics, aſtronomy, and geography, than in chymical or mineralogical knowledge, yet all deſirous of tracing the origin of the globe, and applying Geology to their ſeveral ſyſtems. Thus COSMOGONY was grafted on Geology.

Among many viſionary theories, moſt of which have now ſunk into oblivion, I ſhall ſingle out one as claiming ſome notice from its artful ſtructure and deceptive appearance of ſolidity.

In the formation of this theory, *Genius* (I mean Genius in its primitive ſenſe, the ſublime talent of faſcinating *Invention*, and not the energetic power of patient, profound, and ſagacious *Inveſtigation)* unhappily preſided. Yet dazzled by the ſplendid but delufive fcenery prefented by an ardent imagination foaring to the fource of light, and rending from its flaming orb the planetary maffes that furround it, then marking with daring and overweening confidence fancied fucceffive epochs of the confolidated fabrick of the terraqueous globe, the public attention was long arrefted by the magic reprefentation, and the underftanding nearly betrayed into a partial, if not a total affent to it.

On examining this theory more ferioufly, we may obferve, that two facts, both of which were erroneoufly extended by analogy, and blended with a few inconteftable truths, formed the bafis of all that was not *purely* imaginary in its ftately ftructure.

It was well known that flinty or filiceous fubftances entered into the compofition of common glafs, and alfo that fuch fubftances form one of the principal conftituent parts of the globe; it is alfo certain that many of thefe fubftances refemble glafs in colour, tranfparency, luftre, hardnefs, and fpecific gravity; and as glafs originates from fufion in a ftrong heat, it was thence inferred that thefe fubftances alfo derived their origin from a ſimilar fuſion in an intenſe heat; and as in our planetary ſyſtem ſuch heat could be ſuſpected only in the SUN, it was concluded that theſe ſubſtances were produced *in*, or rather formed a *part* of that luminary, from which, they together with the other planets were detached by the fortuitous ſhock of a COMET. Our globe thus originating, required, it was ſaid, many thouſand years to cool to ſuch a degree as to allow the vapours that accompanied it to condenſe into water, and this water was ſo abundant as to cover it to the height of ſome miles; excavations however were at length formed, into which this liquid gradually ſunk; hence the origin of our *ſeas* and *oceans*. Organic particles (of undefined origin) uniting by *unknown* plaſtic power, peopled the ſeas with *ſhellfiſh*, and ſtocked the continents, when ſufficiently cooled to ſupport them, with *land animals*.

Farther, as ſhells are known to conſiſt of an earth ſimilar to that which forms limeſtone or marble, it was inferred, that after a ſeries of ages, the immenſe maſſes of calcareous ſubſtances, whether found in plains, or forming mountains, originated from accumulated comminuted ſhells, the remains or traces of which are found in many of them at this day.

This proud gigantic theory was, however, like another Goliath, ſoon demoliſhed by a common flint or pebble, the very ſubſtance it ſprung from; common glaſs, eſſentially contains an alkaline ſalt to which alone it owes its fuſibility; ſiliceous ſubſtances contain none, and are abſolutely infuſible when unaſſociated with any. Maquer found them infuſible not only in furnaces, but in the ſtill incomparably ſuperior heat of the concentrated ſolar rays, as did Geyer, Lavoiſier, and Ehrman, in the again higher heat of inflamed oxygen. Hence the hypotheſis grounded on the aſſumed identity of theſe ſubſtances and common glaſs, vaniſhed like the unembodied viſions of the night. With reſpect to limeſtone, the other pillar on which this theory reſted, Cronſted, Ferber, Born, Arduino, and Bergman, demonſtrated the exiſtence of numerous and immenſe mountains, in which not only no veſtiges of ſhells could be traced, but whoſe internal ſtructure or poſition were incompatible with the ſuppoſition of an origination thence derived.

This gaudy illuſion being diſſipated, the internal

internal ſtructure of the globe was more patiently and ſoberly inveſtigated. Mountains offering their conſtituent materials more conſpicuouſly to view, were generally viſited and principally conſulted; the chymical properties, and diſtinctive characters, of their component maſſes, their external relations, whether of poſition, ſuperpoſition, form, direction, or extenſion, the connexion of a peculiar conſtitution, with the abſence or preſence of metallic, or other valuable ſubſtances, and finally, the preciſe height of the whole over the level of the ſea, were now carefully attended to, recorded and publiſhed in moſt parts of the civilized world.

In this magnificent diſplay of the internal arrangement of the globe, many philoſophical obſervers acquired diſtinguiſhed eminence from tedious, laborious, painful, but ſucceſsful, exertions. TILAS, GMELIN, CRONSTED, FERBER, PALLAS, CHARPENTIER, BORN, WERNER, ARDUINO, DE LUC, SAUSSURE, and DOLOMIEU, are names conſecrated to IMMORTALITY; to which I ſhould be proud to add that of a diſtinguiſhed young countryman of our own who has lately travelled into the eaſt, if if his modeſty had not as yet prevented the publication of his obſervations.

I am ſorry to add, that in this highly intereſting career of inquiry, of the firſt importance to civilized ſociety, and demanding the moſt powerful co-operation, few POTENTATES except our own auguſt Monarch, the late amiable ſovereign of France, and the illuſtrious female ſovereigns of Ruſſia, particularly the immortal benefactreſs of mankind, CATHARINE THE GREAT, have taken any concern*. No enterprize however can aſſuredly reflect more laſting ſplendour on their reigns, or more effectually promote the repoſe of their ſubjects, by attracting, abſorbing, and even exhauſting the activity of fiery ſpirits whoſe energies may otherwiſe be perniciouſly employed in diſturbing the conſtitution of the ſtate, inſtead of exploring that of nature †.

As

* The Prince of Brazil may alſo participate of the ſame honourable mention, having, as I am informed, lately ſent to explore the natural hiſtory of that extenſive region. Portugal, in Signor Camera, now poſſeſſes real mineralogical ability.

† Perhaps the ſhare which ſome noted ſcientific men have lately taken in the convulſions of a neighbouring country, may ſeem to invalidate the above aſſertion,

but

As no work comprifing the generality of this object has as yet appeared in the Englifh language, though feveral valuable fragments bearing fome relation to it may be noticed, I thought I fhould make no unacceptable prefent to the public, by collecting and exhibiting to its view, the moft important obfervations that occur in the works of the celebrated writers I have already mentioned, and of many others lefs generally known; the whole connected by fuch theories as appeared to me moft probable, either devifed by others, or refulting from my own reflections. COSMOGONY, an object, confidered in its totality, above the reach of human underftanding, I fhould have avoided

but it fhould be remarked, that of the votaries to *natural knowledge*, many became the victims of that direful tyranny, fuffering either *death* as Lavoifier and Diedrich, or *exile* as Bournon, De Mazieux, La Peyroufe, &c. The few whofe names ftill remain enrolled in the ever execrable annals of anarchy, were neverthelefs in reality guiltlefs of its enormities, being reftrained from oppofition, by the then all prevailing terror. The pretended philofophic reformers of *metaphyfics*, *morality*, and *politics*, and the frantic enemies of chriftianity *alone* prepared, excited, and acted thofe atrocious tragedies, in comparifon of which the accumulated cruelties of ancient tyrants, and of pagan and chriftian perfecutions, are loft to the fight.

meddling

meddling with, had it not been for the pernicious influence, I obſerved ſome falſe but ingenious ſyſtems of it to have on perſons in other reſpects far from ignorant, and the evident agreement I diſcovered between the account of it given by MOSES and the moſt certain and ſtriking geological obſervations. This work was ready for the preſs in June, 1798, but the confuſion ariſing from the rebellion then raging in Ireland, prevented the impreſſion. An opportunity however occurred of ſending the manuſcript to Germany, where I knew its tranſlation would be attended with notes, which might conſiderably improve the intended ſubſequent publication in Engliſh. The German tranſlation I have lately received, accompanied with many notes, ſome of which are very valuable, being extracted from the Journal of Travels through Peru, by Mr. Helm, publiſhed in Germany in 1798, and containing the only exact mineralogical account extant of that intereſting country; theſe I have inſerted and ſubjoined to the preſent publication, with ſuch other remarks as I thought worthy of notice.

TABLE OF CONTENTS.

Trap

Indurated

Sulphur-

ERRATA.

Page	Line	For	Read
Page 1	Line 4	for *ſhall*	Read *ſhould*
—— 8	laſt line	— *have*	—— *having*
—— 10	line 7	— forms	—— *form*
—— 13	— 17	— that	—— *that*
—— 17	— 17	— all	—— *almoſt all*
—— 25	— 22	. . . .	*Dele* the comma after only
—— 36	— 6	— *in*	Read *at*
—— 37	— 25	— prevails . . .	—— prevail
—— 44	— 16	— Argillaceous	—— Martial argillaceous
—— 113	— 5	— Earths . . .	—— Earth
—— 114	— 1	— Decompoſed	—— Diſſolved
—— 118	— 24	— Requires . .	—— Require
—— 123	— 16	— Montarmiata	—— Montamiata
And for		Peperino . .	—— Piperino
—— 125	— 5	— Lapidiſcence	—— Lapideſcence
—— 127	— 15	— Preadamtic	—— Preadamitic
—— 131	laſt line	— it	—— the concreted maſs
—— 147	— 7	after abſorb . . .	—— moiſture
—— 150	— 3	. . . .	*Dele* Vitriolic
—— —-	— 16	— which in its turn	Read forming a ſalt which in its turn
—— 168	— 25	— thus in 1775	—— thus in Swiſſerland in 1775
—— 440	— 19	— were	—— are
—— 457	— 4	— nor	—— but not
—— 465	— 8	— Cartſbad . .	—— Carlſbad

A few others of ſlight moment are left to the indulgence of the reader.

GEOLO-

GEOLOGICAL ESSAYS.

ON THE PRIMITIVE STATE OF THE GLOBE AND ITS SUBSEQUENT CATASTROPHES.

In the inveſtigation of paſt facts dependent on natural cauſes, certain laws of reaſoning ſhould inviolably be adhered to. The firſt is, that no effect ſhall be attributed to a cauſe whoſe *known* powers are inadequate to its production. The ſecond is, that no cauſe ſhould be adduced whoſe exiſtence is not proved either by actual experience or approved teſtimony. Many natural phenomena have ariſen or do ariſe in times or places ſo diſtant, that well conditioned teſtimony concerning them cannot without manifeſt abſurdity be rejected. Thus the inhabitants of the Northern parts of Europe, who have never felt earthquakes nor ſeen volcanos, muſt nevertheleſs admit,

from

ſrom mere teſtimony, that the firſt *have* been, and that the ſecond *do actually* exiſt.

The third is, that no powers ſhould be aſcribed to an alleged cauſe but thoſe that it is known by actual obſervation to poſſeſs in appropriated circumſtances.

To theſe laws I mean ſtrictly to conform in the ſubſequent inquiry, and on this conformity to reſt its merits. To them I ſhall appeal in examining the various ſyſtems I may have occaſion to mention.

To thoſe who may regard this inquiry as ſuperfluous, and conſider the actual ſtate of the globe as alone entitled to philoſophical attention, I ſhall beg leave to obſerve, that its original ſtate is ſo ſtrictly connected with that which it at preſent exhibits, that the latter cannot be properly underſtood without a retroſpect to the former, as will amply be ſhewn in the ſequel. Moreover recent experience has ſhewn that the obſcurity in which the philoſophical knowledge of this ſtate has hitherto been involved, has proved too favourable to the ſtructure of various ſyſtems of atheiſm or infidelity, as theſe have been in their turn to turbulence and immorality, not to endeavour to diſpel it by all the lights which modern geological reſearches have ſtruck out. Thus it will be found that geology naturally ripens, or (to uſe a mineralogical expreſſion) *graduates* into religion, as this does into morality.

So numerous indeed and ſo luminous have been the more modern geological reſearches, and ſo obviouſly connected with the object we have now in view, that ſince the obſcuration or obliteration of the primitive traditions, ſtrange as it may appear, no period has occurred ſo favourable to the illuſtration of the original ſtate of the globe as the preſent, though ſo far removed from it. At no period has its ſurface been traverſed in ſo many different directions, or its ſhape and extent under its different modifications of earth and water been ſo nearly aſcertained, and the relative denſity of the whole ſo accurately determined, its ſolid conſtituent parts ſo exactly diſtinguiſhed, their mutual relation, both as to poſition and compoſition, ſo clearly traced, or purſued to ſuch conſiderable depths, as within theſe laſt twenty-five years. Neither have the teſtimonies that relate to it been ever ſo critically examined and carefully weighed, nor conſequently ſo well underſtood, as within the latter half of this century.

The introduction of teſtimony into reſearches merely philoſophical has been, I am well aware, objected to by many, but in the preſent caſe the objection evidently originates in inattention to its object. All philoſophical reſearches are grounded either on experiment or obſervation ſingly or jointly, and the conſequences clearly deducible from them. Where recourſe cannot be had to experiment, as in the preſent caſe, there obſervation ſingly muſt be reſorted to, but as objects even of obſervation are not of daily occurrence, and many of them muſt have exiſted at diſtant intervals of time and place, recourſe muſt be had to its *records*, and conſequently to teſtimony. Aſtronomy furniſhes us with a caſe in point. This is a ſcience purely philoſophical, yet aſtronomers have never heſitated

to

to admit the obſervations of an Hipparchus or a Ptolemy. In effect, paſt geological facts being of an hiſtorical nature, all attempts to deduce a complete knowledge of them merely from their ſtill ſubſiſting conſequences, to the excluſion of unexceptionable teſtimony, muſt be deemed as abſurd as that of deducing the hiſtory of ancient Rome ſolely from the medals or other monuments of antiquity it ſtill exhibits, or the ſcattered ruins of its empire, to the excluſion of a Livy, a Salluſt, or a Tacitus. That great changes have taken place on the ſurface of the globe ſince the commencement of its exiſtence, changes that for ſome thouſand years have not been repeated, is allowed on all hands. What then ſhould render theſe facts and the circumſtances attending them unſuſceptible of teſtimony? not ſurely their improbability or diſcrepance with actual obſervation, ſince their reality is confeſſed by all; with reſpect to *ſome* of them I can think of no reaſon but one, and that indeed at the firſt bluſh ſufficiently plauſible, namely, that their exiſtence preceded that of the human ſpecies; this certainly proves that the knowledge of the hiſtorian that relates them (ſuppoſing him to have any) was not as to ſuch facts obtained by human means. But if in a ſeries of facts, diſcovered by an inveſtigation to which the witneſs was an utter ſtranger, an exact agreement with the relation of the hiſtorian be diſcerned, not barely as to the ſubſtance of the facts but even as to the order and ſucceſſion of their exiſtence, in ſuch caſe it muſt be acknowledged that the relation is *true*, let the knowledge of the hiſtorian have been obtained how it may. If its primary ſource cannot be human, it muſt have been ſupernatural, and moſt aſſuredly worthy of credit even in ſuch inſtances as have not as yet been corroborated by obſervation, or perhaps are incapable of ſuch additional proof. Now ſuch an account of the primeval ſtate of the globe and of the principal cataſtrophe it anciently underwent, I am bold to ſay Moſes preſents to us, and I make no doubt of demonſtrating in the following Eſſays.

ESSAY I.

ON THE PRIMEVAL STATE OF THE GLOBE.

THE firſt remarkable fact that preſents itſelf to our notice on conſidering the primitive ſtate of the globe is, that its ſuperficial parts, at leaſt to a certain depth, muſt have originally been in a ſoft or liquid ſtate. This fact is inferred from the ſhape it at preſent exhibits, which, as aſtronomers tell us, is that of a ſpheroid compreſſed at the poles, the polar diameter being found ſeveral miles ſhorter than the equatorial; nor is it at the poles only that this compreſſion is obſervable, but in all the higher degrees of latitude, nearly in proportion to their proximity to the poles. This ſhape it evidently could not aſſume unleſs to a certain depth its ſuperficial parts were in a ſoft or liquid ſtate. Some *geological* obſervations alſo indicate that its component parts, even thoſe that are at preſent the moſt ſolid, were originally in a ſoſt ſtate. Thus in the mountains of Quedliœ and Portfiœllet in Norway, which conſiſt of an argillaceous pudding ſtone, the ſiliceous pebbles it contains are obſerved to be compreſſed to the thickneſs of about $\frac{1}{4}$ of an inch in the lower parts of the mountains, but to increaſe in ſize and roundneſs in proportion as their ſituation is higher.— 1. Bergm. Erde Beſch. 182. and in the *Vivarois* the loweſt ſtrata of primitive limeſtone have been found of the thickneſs of only $\frac{1}{10}$ of an inch*, but in proportion to their heighth in the mountain their thickneſs increaſes, until at its ſummit it arrives to thirty or forty feet. 1. Soulavie, 178. Mr. Ferner made the ſame obſervation in England†, but it is needleſs to inſiſt further on this point, as it is now generally allowed.

With reſpect to the interior and more central parts, they have been hitherto in-

* This, however, does not always happen, 1 Sauſſ. 195, for the reaſon given ibid. 453.

† 1 Roz. 8vo. p. 64. and Gruber in Carniola. Phy. Arbeit. 2d. Stuck. 3.

acceſſible,

accessible, nothing can be determined from immediate observation, but we may collect with sufficient certainty, and it is now generally acknowledged, that at the time of the creation, and for many centuries after, they contained immense empty caverns, and consequently consisted of materials sufficiently solid to resist the pressure of the enormous mass of liquid substance placed over them. See Boscovich, and 4 La Metherie 15.

The liquidity thus proved to exist in the more superficial parts of the globe, comprehending even those that are now most solid, must have proceeded either from igneous fusion or solution in water. The hypothesis of igneous fusion wars with every notion which experiment has taught us to form either of fire or its fuel, or the properties and appearances of the various substances supposed to be subjected to it, as I have shewn at large in a former dissertation contained in the Transactions of the Irish Academy; the latter perfectly accords, and much more perfectly than I was then aware of, with all the properties and characters that

that all the ſolids now known exhibit, thoſe confeſſedly of volcanic origin ſolely excepted.

The difficult ſolubility in water of moſt of the ſolids which the globe at preſent exhibits, and the immenſe quantity of that fluid requiſite to effect their ſolution, forms the only difficulty that has hitherto embarraſſed geologiſts, though it has prevented ſcarce any of them from admitting that ſolution. Moſt of them have ſuppoſed that at that early period ſome menſtruum exiſted capable of effecting it. This difficulty, however, proceeds ſolely from inattention to the firſt demonſtrated fact, namely, that the globe at its origin, at leaſt to a certain depth, was a liquid maſs; therefore the ſolids that at preſent compoſe it were not originally in a ſolid ſtate, whoſe converſion into a ſtate of liquidity would certainly require more water than is known to exiſt, but were at the very commencement of their exiſtence in that ſtate of minute diviſion which aqueous ſolution requires, but which no known exiſting quantity of tnat fluid would be able to effect.

Now

Now it is a well known chemical fact, that less of any menstruum is requisite to keep a solid substance in solution, at least for a short time, than originally to dissolve it.

Yet if the quantity of aqueous fluid requisite even to keep the mass of solids in solution were too small, as possibly it may have been, this would only hasten the *second* general fact, to which I now proceed, namely, the crystallization, precipitation, and deposition of these solids. But before I enter on this event it will be necessary to consider more particularly the state of this original chaotic fluid.

The water which constituted this menstruum, being in a liquid state, must have been heated at least to thirty-three degrees, and possibly much higher. Secondly, it contained the eight generic earths, all the metallic and semi-metallic substances now known, the various simple saline substances, and the whole tribe of inflammables, solid, and liquid, which are of a simple nature, variously distributed, forming upon the whole a more complex menstruum than any that has since existed, and con-

sequently

ſequently endued with properties very diſ-ferent from any with which we have been ſince acquainted.

Hence elementary fire or the principle of heat muſt have been coeval with the crea-tion of matter, and the general properties of gravitation and elective attraction may be ſuppoſed of equal date.

The proportion of the different materials contained in the chaotic fluid to each other, may be ſuppoſed upon the whole nearly the ſame as that which they at preſent bear to each other, the ſiliceous earth being by far the moſt copious, next to that, the fer-ruginous, then the argillaceous and calca-reous, laſtly, the magneſian, barytic, Scot-tiſh, and Iargonic, in the order in which they are named, the metallic ſubſtances (except iron) moſt ſparingly; in particular parts, however, of this polygenous fluid a very different proportion muſt have ob-tained (as in ſome parts of the globe) ſome ſpecies of earth or metal, &c. have ever-more been found more copious than in others. Some geologiſts, as Buffon, and of late Dr. Hutton, have excluded calca-

reous

reous earth from the number of the primeval, afferting the maffes of it we at prefent behold, to proceed from fhell fifh. But in addition to the unfounded fuppofition, that fhell fifh or any animals poffefs the power of producing any fimple earth, thefe philofophers fhould have confidered that before the exiftence of any fifh the ftony maffes that inclofe the bafon of the fea muft have exifted, and among thefe there is none in which calcareous earth is not found. Of this circumftance indeed Buffon was ignorant, the analyfes that prove it being unknown to him. Dr. Hutton endeavours to evade this argument by fuppofing the world we now inhabit to have arifen from the ruins and fragments of an anterior, and that of another ftill prior, without pointing at any original. If we are thus to proceed *in infinitum* I fhall not pretend to follow him, but if he ftops any where, unlefs he alfo fuppofes his primitive globe abfolutely different from that which we inhabit, (and with fuch I do not meddle) he will find the fame argument equally to occur.

In a fluid conftituted as that juft mentioned, it is evident from the laws of

elective

elective attraction, that the various solids diffused through it must soon have coalesced in various proportions according to the laws of this attraction and the presence or proximity of the ingredients, and thus have crystallized into different groups, which descended to and were deposited on the inferior solid kernel of the globe *. In those tracts in which the siliceous, and next to it the argillaceous earth most abounded, (and such tracts appear to have been by far the most extensive) granite and gneiss appear to have been first formed, and their formation may thus be explained: Both these rocks consist of quartz, felspar and mica, in a variable proportion, but the quartz and felspar are generally the most copious. These stones are themselves composed of siliceous and argillaceous particles, and particularly the first, principally of siliceous, the two latter admitting also the argillaceous and a small proportion of the calcareous, the magnesian and in some instances of the barytic. Now of these earths, *that* should coalesce first, which

* Mr. De Luc's opinion that stones were formed by mere deposition, is refuted by Charpentier. Saxony, p. 304, 305.

with

with an equally ſmall affinity to water was at the ſame time moſt plentifully contained in it, its particles being more within the reach of each other's attraction. Hence we may conclude that the quartz firſt cryſtallized, ſcarce ever indeed, perfectly, from the diſturbance that muſt have occurred in ſuch an immenſe body of an heterogeneous fluid, nor perfectly pure by reaſon of its affinity to argil and calx*. Next to this felſpar, containing a ſmaller proportion of ſilex and a larger of the other earths, proportions which, from their eaſy fuſibility, alſo appear to exhibit the *maximum* of attraction of theſe earths to each other when ſilex and argil prevail, muſt have cryſtallized next, and laſtly the mica, a ſtone in which the proportion of ſilex to argil is ſtill ſmaller. The portion of water diſengaged from theſe earths gradually aſcended and made room for new ſhoots, which attaining the foregoing before they were perfectly hardened, adhered to them cloſely, and thus at laſt vaſt uniform blocks were formed; where the ſhoots had not attained a certain degree

* Lime water precipitates Silex, from Liquor Silicum, 22 An. Chy. 110. See 6 Sauſſure, 186.

of

of hardneſs, or the ſofter ingredient, viz. the mica abounded, the gneiſs was formed; and where the proportions requiſite to form felſpar were deficient, the other granitic ingredients being preſent, ſhiſtoſe mica was formed. Hence we may underſtand how it happened that gneiſs ſhould ſometimes be found in granite*, and ſometimes maſſes of granite in the midſt of gneiſs, and why in mountains, granite, gneiſs and ſhiſtoſe mica frequently alternate with each other. Charpent. 390†.

As the fluid from which theſe cryſtallized granitic maſſes ſubſided was of the moſt heterogeneous kind, it is not to be wondered at that various metallic ſubſtances, and particularly iron, and even ſome traces of carbon and plumbago, ſhould ſometimes occur in them. 2 Sauſſ. 451. 2 Bergm. Jour. 1790. 532.

In other tracts where the ſame earths occurred, but not in the proportions fitted to produce granitic ingredients, other maſſes of the ſiliceous genus, as ſiliceous ſhiſtus, ſili-

* Werner kurze Claſſif. 9, 10. 6 Sauſſ. 195.

† He ſhews them to be coeval, p. 396 and ante, and 8 Sauſſ. 55.

ceous porphyries, jaſpers, &c. were formed by a leſs perfect and confuſed cryſtallization, or partly cryſtallized and partly depoſited, as in 6 Sauſſ. 128, § 1574.

In various places argillites, hornblende ſlates, ſerpentines and other primeval ſtones of various denominations muſt have ariſen according to the predominant proportion of their ingredients by a more or leſs perfect or partial cryſtallization. 4 La Metherie, 86.

Metallic ſubſtances (all of which I ſuppoſe to have originally exiſted in their complete metallic ſtate) and particularly iron, of all others the moſt copious, muſt in ſundry inſtances have met and combined with ſulphur, the ſubſtance to which all, and particularly iron, have the greateſt affinity, and thus pyritous ſubſtances and ſulphurated ores originated. Petrol, ſpecifically lighter indeed than water, but involved in the chaotic fluid, meeting ſulphur to which it has an affinity, with it formed a liquid ſpecifically heavier, which gradually involved and was abſorbed by carbonic particles which were thus collected and precipitated.

It is a fact at preſent well eſtabliſhed,

that in the act of cryſtallization a very conſiderable degree of heat is generated. Judging by analogy, how great then muſt have been the heat produced by the cryſtallization of ſuch immenſe quantities of ſtony maſſes as took place at this period? the immediate effect of which muſt have been an enormous and univerſal evaporation, ſweeping over the ſurface of the heated fluid according to the inequality of its diffuſion and of the cauſes that produced it in various tracts.

The heat thus produced muſt have been ſtill farther increaſed in conſequence of an event which naturally reſulted from the degree at firſt excited. For in conſequence of the heat and evaporation, the quantity of the chaotic fluid (the univerſal menſtruum), as alſo its ſpecific gravity, were diminiſhed, and thus the ſubſtances contained in it (of which it was not the moſt natural ſolvent) were ſtill more diſpoſed to precipitation, as uſually happens in ſuch caſes; thus then the ferruginous particles naturally not ſoluble while in their metallic ſtate in any fluid, and of which immenſe quantities exiſted, were rapidly and copiouſly precipitated;

ed; the aqueous particles intercepted between them muſt in that caſe have been decompoſed, and an immenſe quantity of inflammable air ſet looſe, the heat thus produced increaſing with the maſſes operated upon, muſt have riſen at laſt to incandeſcence; in that circumſtance the oxygen abſorbed muſt have been in great meaſure expelled, and in its naſcent ſtate meeting and uniting with the inflammable air muſt have burſt into flame. The progreſs of ſuch high degrees of heat muſt have diſengaged all the oxygen contained in the contiguous chaotic fluid, which uniting partly with more metallic iron, partly with the ſulphurated and partly with the carbonic and bituminous ſubſtances, muſt have occaſioned a ſtupendous conflagration, the effects of which may well be ſuppoſed to have extended even to the ſolid baſis on which the chaotic fluid repoſed, and to have rent and ſplit it to an unknown extent.

That flame ſhould thus burſt from the boſom of the deep is not a forced ſuppoſition, but has frequently been verified in latter times. I ſhall only mention one inſtance which happened in the beginning of this century, when flames burſt out of the ſea near Tercera, and an iſland was elevated*.

Theſe volcanic eruptions, many of which ſeem to have taken place at this period, chiefly in the ſouthern hemiſphere, were attended with important conſequences; the firſt muſt have been the diffuſion of a conſiderable heat through the whole maſs of the chaotic fluid, by which means the oxygen and mephitic airs diſperſed through it muſt have been extricated, and thus gradually formed the *atmoſphere*.

The ſecond was the production of fixed air from the union of oxygen with the ignited carbon; this at firſt roſe into and diffuſed itſelf through the atmoſphere, but in proportion as the chaotic fluid cooled it was gradually abſorbed by it. This abſorbtion occaſioned the precipitation, and more or leſs regular cryſtallization of the *calcareous* earth, the greater part of which being much more ſoluble than the other earths,

* 6 Phil. Tranſ. Abridg. 2d Part, 203.

ſtill

ſtill remained in ſolution after the others had, for the moſt part, been depoſited. This explains why many of theſe primitive calcareous maſſes are of all others the freeſt from foreign admixture. In ſome inſtances, however, it muſt have happened that the calcareous particles intermixed with other earths were ſaturated before the reunion and depoſition of the other earths, and hence, in ſome countries, ſtrata of primitive calcareous maſſes occur in the midſt of gneiſs, or alternating with it; but theſe inſtances are very rare *.

That the formation of fixed air was an event ſubſequent to the formation of moſt of the primeval ſtones, appears from the obſervation, that the calcareous earth found in the compoſition of primeval ſtones is in a cauſtic ſtate. 44 Roz. 206.

The immenſe maſſes concreted and depoſited on the interior nucleus of the earth formed the *primitive mountains*. It may, perhaps, be thought that this depoſition ſhould be equally diffuſed, and ſhould conſtitute only an even cruſt over this interior nucleus, but ſuch a diſpoſition is contrary

* See Charpentier, 399, 402, 403.

to the nature of cryſtallization, between whoſe ſhoots an interval always intercedes, if not too ſudden. The water firſt diſcharged of its diſſolved contents, and thereby heated, moved upwards, being preſſed by the circumambient denſer fluid, which was too heterogeneous to mix with, and be diluted by it. The depoſit already formed affording to the ſucceeding portions of the charged fluid a baſis whoſe points of contact were ſo much the more numerous as its height was greater, thereby determined theſe portions to a ſimilar depoſition, until the diminiſhed denſity or exhauſtion of the menſtruum diminiſhed or put an end to the number and extent of the points of contact of the depoſited maſſes with the ſolids contained in the menſtruum. In ſome caſes alſo, particularly after the chaotic fluid was heated by the cauſes already mentioned, and a conſiderable evaporation had enſued, the cryſtallization might have begun at the ſurface, as we ſee happen to ſome ſalts and to lime-water. Thus extenſive ſtrata might have been ſucceſſively depoſited, moſtly in an horizontal, but often

ten from accidental ruptures during their fall, in an oblique, or nearly vertical, position. Thus far we are led by general analogies, without the aſſiſtance of romantic or gratuitous hypotheſes, and the view of the ſubject, thus obtained, is ſufficient for the explanation of moſt of the obſervations hitherto made on the ſtructure of primitive mountains.

The formation of *plains* is eaſily underſtood; in the wide intervals of diſtant mountains, after the firſt cryſtallized maſſes had been depoſited, the ſolid particles ſtill contained in the chaotic fluid, but too diſtant from each other's ſphere of attraction to concrete into cryſtals, and particularly thoſe that are known to be leaſt diſpoſed to cryſtallize, and alſo to have leaſt affinity to water, were gradually and uniformly depoſited. Of this nature argillaceous particles are known to be, intermixed as may well be expected with a large proportion of ſiliceous and ferruginous particles of all others the moſt abundant, and ſome particles of the other earths; by theſe compound and ſlightly concreted earths the ſurface of plains were originally covered. In process of time these earths undoubtedly received an abundant increase from the decomposition of primitive mountains, but this being an event of a posterior date need here be only cursorily mentioned.

The next important event necessary to fit the globe for the reception of land animals, was the diminution and recess of the chaotic fluid in whose bosom the mountains were formed, and the consequent disclosure of the dry land. This event was the natural consequence of the operation of the preceding volcanoes, by these the bed of the ocean was scooped, most probably as we shall hereafter see, in the *Southern* hemisphere. But no change or transposition of the solid materials deposited from the chaotic fluid could lower its level, unless the inferior *nucleus* of the globe could receive it within its hollow and empty caverns; this admittance it gained through the numerous rifts occasioned by the antecedent fires; at first rapidly, but afterwards more slowly, in proportion as the perpendicular height of the fluid was diminished, and thus the

emerged

emerged continent conſiſting of mountains and plains was gradually laid bare and dried, and by drying, conſolidated.

The diſcloſure of the actual continents, as I have juſt hinted, appears to have been gradual. The tracts at firſt uncovered were thoſe whoſe height over the preſent ſeas amounts to 8500 or 9000 feet, or more. This height comprehends moſt of the Eaſtern heights of Siberia, between latitude 49° and 55°, and of the extenſive regions of Great Tartary, Thibet, the deſert of Coby or Chamo, and China, reaching in ſome places to latitude 35°, and extending in the Northern parts from the ſources of the Irtiſh, long. 95°, and in the more Southern from the heads of the Ganges and Bourampooter, long. 80°, Hohanho and Porentſho to long. 190° at the leaſt, and, perhaps, ſtill farther into the unknown parts of Eaſtern America.

In Europe, only, the ſummits of the Alps, Pyrenees, and of a few other mountains, were uncovered, but in America the narrow but long chain of the Cordeliers muſt have raiſed its ſummits far above the ocean;

ocean; this fact is a neceſſary inference from that which I am next to mention.

The level of the ancient ocean being lowered to the height of about 8500 or 9000 feet, then, and not before, it began to be peopled with *fiſh*. I ſhall, therefore, from its ſimilitude to our preſent *ſeas* henceforward denote it by this denomination, to diſtinguiſh it from the *chaotic fluid*, whoſe compoſition was ſo different and contained no fiſh.

That the creation of fiſh was an event ſubſequent to the emerſion of the tracts juſt mentioned and to the reduction of the waters to the height I have ſtated, is proved by the obſervations of all thoſe who have viſited thoſe countries. Pallas informs us that the immenſe deſert of Cobea or Chamo forms a flat platform whoſe elevation can be compared only to that of Quito in Peru (which Bouguer has ſhewn to amount to upwards of 9000 feet), and that the *plains* of the Moguls all along to the Chineſe wall are nearly of the ſame height*.

* 1 Act. Petrop. 1777, p. 38.

Major Rennel, in his account of the map of India, tells us, that the country of Thibet is one of the higheſt in Aſia, being part of that elevated tract which gives riſe not only to the rivers of India and China, but to thoſe of Siberia and Tartary. The Southern ridge of the mountains of Boutan (Thibet) riſes, he ſays, about a mile and a half above the plains of Bengal, and may be ſeen in a horizontal diſtance of one hundred and fifty miles, p. 93 and 94, which indicates a height of fifteen thouſand feet, or allowing for refraction according to Dr. Maſkelyne's rule, eleven thouſand nine hundred and eighty-one feet *.

According to Abbé Man's calculations, of

* The Mountains of Chineſe Tartary, N.E. of the Chineſe wall, have been found, by barometrical meaſurement, to be 15810 feet, 1 Bergm. Erdekuzel 172. If the miſtake amounted even to 3000 feet, which is the moſt that can be ſuppoſed, ſtill their height would exceed 12000 feet.

The aſcent to Tartary is ſuch, that ſome parts of it have been aſcertained to be 15000 feet above the ſurface of the Yellow ſea, 2 Staunton 206. See alſo p. 508. The mountain that divides Kiangſee province from that of Chantung, the ſource of the river, is about 7000 feet high.

of all others the moſt accurate and the moſt moderate, the height of the Ganges and Hohanho, even at one thouſand miles diſtance from the ſea, muſt be three thouſand ſix hundred and thirty feet *. But Major Rennel has ſhewn in the Philoſophical Tranſactions, 1781, p. 90, that the head of the Ganges is two thouſand miles diſtant from the ſea, therefore the country it flows from is elevated at leaſt eight thouſand eight hundred, or nine thouſand feet.

Now in theſe elevated tracts no marine ſhells or petrifactions are found in the body of any mountain, nor in any ſtone, not even in limeſtone, though it abound particularly about the ſources of the Amour, Herm. 1 Chy. An. 1791, p. 155. But all the calcareous maſſes that occur are either what are called *ſaline* like Carrara marble, or ſo fine grained foliated as to appear nearly compact, but of the primitive kind. This abſence of marine ſhells and petrifactions from ſuch extenſive regions has attracted the particular notice of all travellers into theſe parts as they are ſo abundantly found in all lower tracts of the globe, Gmelin, 45.

* Phil. Tranſ. 1779.

Phil.

Phil. Tranſ. 254. Pallas, 1 Act. Petrop. 44. Patrin, 38 Roz. 227. And though ſalt ſprings and lakes are found in the higheſt plains, Pallas, ibid. 38, and even coal mines in the mountains, yet no organic remains accompany theſe mines as they do in the lower tracts of the globe, Patrin, 38 Roz. 226. Pallas indeed remarks ſome few petrifactions have been found in the rifts even of granitic mounts, but theſe he rightly judges were depoſited there at the time of the deluge; ibid. 44.

Hence I think it follows evidently that theſe tracts were indeed formed in the boſom of the primitive ocean, like all others, but that they were uncovered before the creation of fiſh, and ſince they contain limeſtone, that this ſtone does not neceſſarily and univerſally originate from comminuted ſhells, as Buffon and others have advanced *.

That fiſh did not exiſt until the level of the ocean was depreſſed to about eight thouſand five hundred feet, may alſo be inferred with

* See alſo Charpentier's Demonſtration of Buffon's Error, p. 399 and 402, &c.

equal

equal evidence from this obſervation, that though ſeveral lofty mountains at preſent exiſt, which far exceed that height, yet no petrifactions or ſhells are incorporated in the rocks or ſtrata that form them *. This De la Peyrouſe atteſts with reſpect to the Pyrenees, which yet are moſtly calcareous. *Traite des Mines de fer*, 336. Nor are any found in *Santo Velino*, the higheſt of the Appenines, its height being eight thouſand three hundred feet, whereas they abound in thoſe that are lower. In the Savoyan Alps, Salenche, Saleve, Mole, the Dole, all of which are calcareous, but below the height of ſeven thouſand feet, contain petrifactions, but the Buet, which is alſo calcareous for the greater part, but whoſe height exceeds ten thouſand feet, contains none. Sauſſure paſſim. But quere if the Buet be not ſecondary? 7 Sauſſ. 296. and 1

* Shell fiſh appear to be of all others the moſt ancient; perhaps the reaſon might be, that they could live in water more turbid with heterogeneous ingredients, and more fouled with Petrol, than other fiſh, or becauſe the ſea was originally more ſalt.

Many amphibious marine animals require the exiſtence of land to produce their young. See 38 Roz. 283.

Sauſſ.

Sauff. § 590; it is not, though in vol. 7 he feems to fay it is.

And reciprocally of the mountains that *contain* petrifactions embodied in their mafs, none reaches to the height of eight thoufand fix hundred feet. With refpect to the Hartz, Lafius remarks, that no petrifactions are found in the mountains whofe height exceeds two thoufand three hundred and ten feet, Lafius, 148; and Renovants, p. 76, afferts, that none of thofe of Siberia on either fide of the Altai exceed two thoufand Paris feet. Nay Pallas does not allow to calcareous mountains that contain petrifactions (which on this account he judges to be formed in the fea) above thirteen or fourteen hundred feet, 1 Act. Petrop. 59; but a view of the mountains of Switzerland fhews that many of them are much higher.

The neareft approach I have met with, to the limits I have affigned to the height of the ancient ocean when it began to be peopled with fifh, is an obfervation of Mr. De Luc's, that he found cornua ammonis petrified on mount Grenier, whofe height is feven thoufand eight hundred feet, 2d

Lettres

Lettres à la Reine, 227; and another of Baron Zoits on the mountain of Terglore in Carniola, that petrifactions were found in limestone at the height of between thirteen and fourteen hundred German lachters; taking such to be equal to those of the Hartz, the height was eight thousand five hundred English feet nearly, and hence I have extended the limits to eight thousand five hundred feet, yet it does not appear by Mr. De Luc's relation that these petrifactions were imbodied in the stone or rock which constitutes the mountain, and therefore may well be only relics of the deluge. He also says that the chain of Jura abounds in marine remains; but Jura consists of two chains, the highest of which consists of primeval limestone in which no petrifactions are found, and this is the highest, 30 Roz. 275, and 11 Annales Chy. 265, but the lower chain certainly abounds in them. Fichtel also remarks that still higher on the mountains of Terglore no petrifactions are found imbodied, Mineral Aufsatz, p. 4.

After this elevated tract of the globe had been uncovered, there is no reason to suppose that it long remained divested of vegetables or unpeopled by animals, being in every reſpect fitted to receive them. The ſevere degree of cold which at preſent diſtreſſes theſe countries during the winter months, is ſolely owing to their diſtance from, and elevation over, the actual ſeas (as I have elſewhere ſhewn), circumſtances that did not exiſt at this period.

That the retreat of the ſea from the lower parts of our preſent continents was gradual and not fully effected until after the lapſe of ſeveral centuries, many reaſons induce us to believe. 1°. Both ſides of the Altaiſchan platform exhibit ſecondary mountains (ſo I call all thoſe which contain marine ſhells or other remains of animal or vegetable ſubſtances between their ſtrata, or incorporated in the rocks or ſtones of which they conſiſt) both of the calcareous and argillitic kind, in which marine ſhells abound, Renov. 75; theſe therefore muſt have been formed before the ſea had receded far from theſe parts, and muſt have been the work of many years.

2do. Not only in every region of Europe, but alſo of both the old and new continents,

D whoſe

whoſe ſituation is inferior to that above mentioned, immenſe quantities of marine ſhells either diſperſed or collected have been diſcovered. In the province of Touraine in France, at one hundred miles diſtance from the ſea, there exiſts at a depth of eight or nine feet a heap of ſhells of nine leagues in ſurface, and upwards of twenty feet in depth, many of which are thoſe of the neighbouring ſeas, a collection which certainly required many years to accumulate. —Mem. Paris, 1720. p. 524, 540. Moſt of theſe ſhells are placed on their flat and not on their convex ſurface—Ibid. which ſhews they muſt have been gently depoſited, and not huddled together by a ſudden and violent inundation. In ſome places ſhells of different ſpecies are thus accumulated, but in others they are regularly arranged in families.—1 Bergm. Erdekug. 251. 262. 6 Roz. 120. Widem. Verwandl. 118. Mem. Par. 1747. 1059. which ſhews alſo that they were neither ſuddenly nor promiſcuouſly collected. Many ſmaller but ſimilar accumulations occur in England, as may be ſeen in the Philoſ. Tranſ. and Ray's Diſcourſes,

Diſcourſes, and in Peru 2 Don Ulloa's Voy. 197. and alſo in Italy, Spain, Germany, Poland, Sweden, &c. which being generally known and acknowledged it is needleſs to detail, but it deſerves particular attention, though many of them are found at a depth of from eight to one or two hundred feet under the ſurface of the earth, and at ſtill greater depths from the ſurface of mountains, yet ſcarce any are found lower than the actual ſurface of the bed of the ſea.—1 Bergm. Erde, 176. 2 Wms. 183, contra, but this is in the Atlantic, a ſea newly formed. Some indeed are found that do not now occur in the neighbouring coaſt, becauſe they are of the kind called *Pelagicæ*, which exiſt only at great depths, as Don Ulloa has ſhewn, 2 Voy. 197. or becauſe the temperature of the actual ſeas is unfit for them, as the ancient ſeas, from the ſmaller extent of the ancient continents, muſt have been much warmer; nay the number of theſe unknown ſhells is daily diminiſhing.

We may alſo remark that the loweſt countries, as Brabant and Holland, contain moſt of them. In particular tracts of Ruſſia they are alſo very copious, becauſe theſe tracts were longer covered by the ſea than moſt other countries, as will be ſhewn in a future Eſſay.

4°. Trees of different kinds and various vegetables have been found in great depths in our modern continents, and even under lofty mountains, as at Meiſſen in Heſſia, and often mixed with marine remains. Some parts of the earth on which theſe trees grew muſt therefore have been dry land, while thoſe parts in which the trees are found were covered by the ſea, therefore the retreat of the ſea muſt have been gradual. Many I am ſenſible aſcribe theſe depoſitions to the ravages of the univerſal deluge, and in ſome caſes, I believe, juſtly, but as they are often under hills or mountains, whoſe ſtrata are regularly diſpoſed, a regularity that can ſcarce be ſuppoſed to take place during the turbulence of a deluge, in theſe caſes it appears their depoſition muſt be aſcribed to more tranquil cauſes.

5°. Trees have often been found depoſited near the ſummits of many mountains at heights in which from the degree of cold which

which at preſent prevails on them they could not grow, therefore they muſt have grown when the temperature of theſe ſummits was warmer, and conſequently when they were leſs elevated over the ſurface of the ſea, and leſs diſtant from it. See the Note in 1 Berg. Erdekugel, 253 of the Pholades and De Luc.

Laſtly,—Stratified mountains of various heights beneath eight thouſand feet exiſt in different parts of Europe, and of both continents, in and betwixt whoſe ſtrata various ſubſtances of marine, and ſome vegetables of terreſtrial origin repoſe, either in their natural ſtate or petrified; the regularity and uniformity of theſe ſtrata ſtrongly indicate a cauſe whoſe action was regular and uniform and long continued; now tides are the only cauſe of this nature with which we are acquainted; ſudden and violent inundations and of ſhort continuance are incompatible with ſtratifications ſo regular and numerous. In ſome few inſtances, it is true, much irregularity and confuſion prevails amongſt the ſtrata, from the diſſolution, elapſion, or different compreſſion of ſome of theſe ſtrata, or from their inter-

 ruption

ruption during their formation, by the introduction of extraneous rocks projected amongst them by adventitious causes, as earthquakes or volcanoes; or the tides themselves may have sometimes been rendered irregular by storms, currents or the irregular profile of the shores, as we daily observe. But even in these cases the length of time necessary for the formation of the strata excludes all suspicion of short and tumultuary inundations.

We also observe, that though the strata themselves are not arranged in the order of the specific gravities of the materials that compose them, the lighter being frequently placed beneath the heavier, yet within each stratum the materials that compose it are almost always arranged according to the laws of their gravitation. 1 Bergm. Erde. 179.

These phenomena fully prove that the retreat of the sea from the vicinity of the lofty mountains and elevated platforms that first emerged was not sudden, but continued for several ages, and that the various stratified secondary mountains at present existing were formed within it during its re-

treat,

treat, and after the creation of fiſh. The mode of their formation now claims our attention.

To form as juſt an idea of the formation of ſecondary mountains as the nature of an object inſcrutable to human eyes can allow us *, we muſt obſerve, 1°. That the greater part of the particles of ſolid matter contained in the chaotic fluid being depoſited before the creation of fiſh, the various materials that enter into the compoſition of ſecondary mountains muſt have been furniſhed either by the deſtruction of ſuch of the primary as exiſted in the ſea, but either from want of ſolidity or the ſmallneſs of their maſs were too feeble to reſiſt its impetuoſity when animated by ſtorms, and being by continued friction reduced to atoms, or rolled into tumblers, were either diffuſed through, or hurried along by the agitation of the waters, or were crumbled to pieces by earthquakes, and variouſly diſperſed through the ocean, or theſe materials were ejected in immenſe heated maſſes, by ſubmarine volcanoes, into the boſom of the waves, to be by them

* 1 Sauſſ. 529.

farther comminuted, disintegrated, or decomposed.

The various solids thus diffused at different periods of time through the vast body of the ocean, must have been gradually precipitated and deposited on such solid masses as resisted the progressive motion impressed upon the precipitating masses by that tumultuous element; hence they applied to and rested on the low lateral surfaces of many of the most considerable primary mountains, or were accumulated on the scabrous but firmly rooted fragments of such of those mountains as were before destroyed, intombing the shell-fish that adhered to or rested upon these fragments, and arresting by their initial softness the various sunk woods and such other vegetable or animal substances as chanced to be mixed with these precipitating masses, or were subsequently borne upon them. Trees naturally assumed the situation that afforded least resistance to the currents that conveyed them, and hence the uniformity that has been observed very frequently in their position. These depositions, when during their descent they attained a certain degree of denſity, muſt have proved fatal to the various ſpecies of fiſh which were involved in them, and hence the origin of the more ſolid piſcine remains at preſent found in them; the ſofter parts being deſtroyed by putrifaction, in this manner, but after long intervals of time, the ſucceeding ſtrata appear to have been formed, but they did not attain their preſent ſolidity until after the retreat of the ſea, and through the operation of cauſes which I ſhall preſently mention.

From the circumſtances here ſtated, we may eaſily account for ſeveral of the moſt general geological obſervations, as,

1°. Why ſtratified hills have always been found repoſing on primary rocks, or inveſting primary mountains, as on granite, gneiſs, argillite, &c. as Ferber, Born, Pallas, Gerhard, Charpentier, Werner, Sauſſure, &c. atteſt *, not as matters of opinion, but as facts they have conſtantly been witneſſes to. Doctor Hutton, indeed, thinks this obſervation does not hold true with reſpect to Scotland, or even with reſpect to the greater part of Britain, becauſe in traverſ-

* 1 Gerh. Geſch. § 77.

ing

ing the greater part of that iſland he ſeldom met with granite; but it is plain from what has been ſaid, that it is not on the ſurface it ſhould be expected, but under the ſuperior ſtrata, therefore the Doctor's obſervations, but not his aſſertions, are perfectly conſiſtent with thoſe of all other geologiſts. However Mr. Everſman of Berlin, who reſided ſome years in Scotland, tells us that the fundamental rock (*Grund gebirge*) of Scotland is a maſs of the *granitic kind*, 1 Bergm. Jour. 1789, 495; and Dr. Aſh relates that not only near Aberdeen, but in the more ſouthern parts, extenſive granitic mountains often occur, 1 Chy. Ann. 1792, 115. The truth is that the whole of Britain ſeems to have been formed of ſtrata, reſting upon and intercepted between maſſes of granite or other primitive rocks, at depths in moſt places not eaſily acceſſible.

2do. Why the ſtrata of ſecondary mountains are generally elevated in the direction that faces the next primitive mountains, though frequently diſtant from them, being, as we have ſeen originally, formed againſt, and along the ſloping ſides of thoſe mountains,

tains, and the ſeparation occaſioned by the courſe of waters that formed the vallies that intercede between them *. The ſeeming exceptions to this rule ariſe either from the ſinking of the primitive mountain to which the ſecondary faced, as happened to that which ſtood where the lake of Geneva is now placed, whoſe bottom is chiefly granite, or ſome other primitive rock, as Mr. Sauſſure atteſts, 1 Sauſſ. 15, and therefore the elevation of the ſtrata of the Mole, Saleve, &c. face the lake, Id. p. 222, 223, &c. or becauſe theſe ſtratified hills were originally independent, being formed on the fragments of primitive mountains, as already ſaid.

Hence alſo we may explain an important obſervation made by Mr. Schrieber on the mountain of Gardette, which conſiſts of limeſtone ſuperimpoſed on gneiſs; he obſerved that where they joined, the gneiſs had penetrated into the body of the limeſtone, but the limeſtone did not penetrate into that of the gneiſs, whence he juſtly inferred, that the limeſtone was in a ſoft

* 5 Sauſſ. 293.

ſtate,

ſtate, but the gneiſs already conſolidated, when the contact took place; 36 Roz. 359. Similar to this is the obſervation of Mr. Sauſſure, 1 Sauſſ. 528, that a puddingſtone with a calcareous cement or argillaceous grit is generally interpoſed betwixt the uppermoſt ſtrata of the primary and the loweſt of the ſecondary ſtrata as the ſoft ſecondary matter inveloped the *pebbles* or *gravel* on the ſurface of the primary; this he obſerved in the Alps, the Voſges, the Cevennes, &c. This is alſo conſtantly obſerved in coal mines, where ſemi-primigenous ſtrata (Todliegendes) a ſandſtone or breccia, with a calcareous or argillaceous cement, forms the laſt ſtratum immediately over the primitive rock.

Many, indeed moſt, of the obſervations hitherto made, are explicable on theſe principles, which, not to extend this part of the Eſſay beyond its due limits, I here omit.

The retreat of the ſea appears to have continued through the rifts already mentioned, or poſſibly through others ſubſequently made, probably until a few centuries before the deluge. Its ceſſation long before

before the deluge I infer from the hardneſs which the mountains muſt have acquired to withſtand the ſhocks they muſt have underwent during that cataſtrophe. To acquire this hardneſs a long period of time was neceſſary, both for their deſiccation and the infiltration of thoſe particles to which the ſtrata of ſecondary mountains owe their ſolidity.

I do not by this pretend that the ſecondary ſtrata had not acquired a certain degree of ſolidity in a few minutes, even after their depoſition, and conſequently long before their emerſion; on the contrary, this muſt ſoon have been acquired in the ſame manner as we find calcareous depoſits to harden at the bottom of ſalt-pans, and tea-kettles, while full of liquor, and tartar in hogſheads of wine, and pouzzolana mortar, &c. but this hardneſs is moderate in compariſon of that which they acquired by deſiccation and continued infiltration, as we daily experience in moiſt limeſtone quarries, where though the ſtone is originally hard, yet it becomes much harder when dried. The induration effected by infiltration is ſtill more conſiderable, as by it the

minuter

minuter particles of bodies are conveyed into the minutest interstices. Hence we see that traps and basaltic pillars are always harder at the bottom than at the top, Cronst. § 267, and the upper strata of limestone, particularly of that species called freestone, are always softer near the surface of the quarry than at a greater depth; that such infiltration is not an imaginary process, let it be explained how it may, and consequently that the hardness of the lower strata does not proceed merely from the pressure of the upper, appears by an elegant observation of M. Werner's, that where various strata of a different nature occur, the petrifactions that are found in the inferior, are frequently filled with the matter of the superior instead of that of the stratum which contains them, Wedem. Umwandl. 118. The petrified shells found in clay or argillites are commonly compressed and flatted, as Mr. Bergman remarks, as argil hardens by contraction, but those found in limestone retain their primitive shape, as these harden chiefly by infiltration.

I now proceed to the Mosaic account of these

theſe events, " In the beginning God cre- " ated the heaven and the earth," that is to ſay, the firſt event in the hiſtory of this globe was its creation, and that of all the planets then known.

" And the earth was without form and " void," that is to ſay, that the earth at the time of its creation was without form, &c. therefore another terraqueous globe did not previouſly exiſt in a *complete ſtate* out of the ruins of which the preſent earth was formed, as ſome have lately imagined; *without form and void* the Hebrew has *Tohu* and *Bohu*. Ainſworth remarks that *Tohu* ſignifies a ſtate of confuſion, and *Bohu* a ſtate of vacuity; ſee Pool's Synopſis. That is to ſay, that the earth was partly in a *chaotic* ſtate, and partly full of *empty* cavities, which is exactly the ſtate, which from the conſideration of the ſubſequent phenomena, I have ſhewn to have been neceſſarily its primordial ſtate.

" And darkneſs was on the face of the " deep," conſequently light did not at firſt exiſt. The *deep*, or abyſs, properly denotes an immenſe depth of water, but here it ſignifies, as Mede and Eſtius obſerve, the mixed

mixed or chaotic maſs of earth and water.—David, whoſe knowledge was derived from Moſes, and who probably poſſeſſed a leſs abridged copy of Geneſis than we do, expreſsly tells us that the earth was covered with water; " the abyſs, like a garment, " was its covering." Pſalm civ. v. 6.

Hence we ſee that the water was from the beginning in a liquid ſtate (and not in that of ice) as I have mentioned; and conſequently elementary fire, or the principle of heat, exiſted from the beginning.

" And the ſpirit of God (or rather *a* ſpi-" rit of God) moved on the face of the wa-" ters;" here *ſpirit* denotes an inviſible elaſtic fluid, viz. the great *evaporation* that took place ſoon after the creation, as ſoon as the ſolids began to cryſtallize, as I have ſhewn. *Of God*, is a well known Hebrew idiom, denoting *great*; *moved*, or rather *hovered*, over the waters. David here mentions a fact which he undoubtedly took from Moſes, though omitted in our preſent copies of Geneſis, and this fact is eſſential to our theory, namely, that " the waters ſtood *above the mountains*." Pſalm civ. v. 6. Therefore the mountains were formed in the

the boſom of the waters, as I have ſtated. Nay, he uſes an expreſſion that moſt probably hath hitherto been ill underſtood, that "God fixed the earth on its baſis, from "which it ſhall not be removed for ever." This appears to me to denote the depoſition of the ſolids contained in the chaotic waters, on the ſolid kernel of the globe, from whence they ſhould never be removed, nor indeed have they ever ſince.

"And God ſaid, Let there be light, and "there was light;" here we may obſerve that *facts* only, were revealed to Moſes or the perſon (moſt probably Adam) from whom their tradition deſcended. The words choſen by Moſes, or this perſon, were ſuch as coincided with his own notions, or were moſt intelligible to an ignorant people. The phraſes, *God ſaid, God ſaw it was good, God called,* uſed in this chapter, are mere anthropological phraſes, ſuited to the conception of thoſe to whom theſe facts were related, for religious and moral, and not merely for ſcientific purpoſes. To men of ſcience their ſignification could not be ambiguous. "*God ſaid,*" ſignifying no more than that events naturally poſſible took place by virtue of the laws of their production, which laws God had established. "*God saw it was good,*" signifies merely that it was good, and the expression "*God called,*" denotes no more than that it received such a name.

The production of light stands next in the order of events recorded by Moses, as it does in our theory, and most probably denotes the *flames* of volcanic eruptions, the Hebrew certainly bears this signification. The period of its existence Moses called *day*, evidently from its resemblance to *true* days, which could have existed only at a subsequent period, namely after the sun had gained its luminous powers.

"And God said, Let there be a firma-
"ment in the midst of the waters, and let
"it divide the waters from the waters."
Here Moses indicates the production of the atmosphere, the word which in our translation is rendered *firmament* * most properly

* Some interpreters think *firmament* implies something solid, because the general opinion of the Heathen oriental sages was, that the heavens were solid, as if the *true sense* of the Hebrew was to be derived from *false opinions* devised many ages after Moses.

signifies

ſignifies *expanſe*, or an expanded or dilated ſubſtance; than which a more proper name could not ſurely be choſen for the atmoſphere. "To divide the waters from the "waters," that is, to ſeparate and contain vapours, which is one principal uſe of the atmoſphere.

"And God ſaid, Let the waters under "the heavens be gathered together in one "place, and let the dry land appear, and it "was ſo." This is the fifth event which Moſes places in the ſame order of ſucceſſion that mere philoſophical conſiderations aſſign to it.

The word *appear* is remarkable, as it ſeems to denote that the diſcloſure of the earth was *ſucceſſive*, and had not from the beginning fully and completely taken place.

The events immediately ſubſequent I omit, as not relating to geology, and ſhall only mention the creation of fiſh, a fact of great importance in the theory of the earth; this, Moſes, as well as philoſophy, tells us happened after the ſeparation of the waters from the dry land and primitive mountains. He alſo relates that the creation of land-animals was subsequent to that of fish; a fact which geological observations also indicate, for their remains are always found near the surface of the earth, whereas those of fish are found at the greatest depths. This order of succession is not only allowed by Mr. Buffon, but made one of the principal pillars of his system. 1 Epoque's de la Nature, p. 231, in 8vo.

Here then we have seven or eight geological facts, related by Moses on the one part, and on the other, deduced solely from the most exact and best verified geological observations, and yet agreeing perfectly with each other, not only in *substance*, but in the order of their succession. On whichever of these we bestow our confidence, its agreement with the other demonstrates the truth of that other. But if we bestow our confidence on *neither*, then their *agreement* must be accounted for. If we attempt this, we shall find the *improbability* that both accounts are false, infinite; consequently one must be true, and, then, so must also the other.

That two accounts derived from sources totally distinct from and independent on each other ſhould agree not only in the ſubſtance but in the order of ſucceſſion of two events only, is already highly improbable, if theſe facts be not true, both ſubſtantially and as to the order of their ſucceſſion. Let this improbability, as to the ſubſtance of the facts, be repreſented only by $\frac{1}{10}$. then the improbability of their agreement as to ſeven events is $\frac{1^{\cdot 7}}{10^{\cdot 7}}$, that is, as one to ten million, and would be much higher if the *order* alſo had entered into the computation.

ESSAY II.

ON THE DELUGE.

Having, I flatter myſelf, eſtabliſhed, in the preceding Eſſay, the credit due to Moſes on mere philoſophic grounds and abſtracting from all theological conſiderations, I ſhall not ſcruple taking him as a guide as far as his teſtimony reaches, in tracing the circumſtances of the moſt horrible cataſtrophe to which the human and all animal ſpecies, and even the terraqueous globe itſelf, had at any period ſince its origin been expoſed. His teſtimony is indeed in ſubſtance confirmed by the traditions of many ancient nations, which may be ſeen in Grotius de Veritate, Huet Queſt. Alnet, lib. ii. chap. xii. Euſeb. Prep. Evang. lib. ix. &c. and therefore needleſs to adduce; it is more to our purpoſe to prove it by geological facts, of which there are ſome that ſeem to me perfectly concluſive.

1ſt, According to Don Ulloa, ſhells were found on a mountain in Peru at the height of of 14220 feet, 2 Buff. Epoque, 268. Mem. Par. 1771, p. 439, in 8vo. 1 Gentil Voy. 116, in 8vo. Now I have already ſhewn, in the former Eſſay, that no mountains higher than 8500 feet were formed ſince the creation of fiſh, or, in other words, that fiſh did not exiſt until the original ocean had ſubſided to the height of eight thouſand five hundred feet above its preſent level. Therefore the ſhells found at more elevated ſtations were left there by a ſubſequent inundation. Now an inundation that reached ſuch heights could not be partial, but muſt have extended over the whole globe.

2dly, The bones of elephants and of rhinoceri, and even the entire carcaſe of a rhinoceros have been found in the lower parts of Siberia. As theſe animals could not live in ſo cold a country, they muſt have been brought thither by an inundation from warmer and very diſtant climates, betwixt which and Siberia mountains above nine thouſand feet high intervene*. It may be replied that Siberia, as we have already ſhewn, was not originally as cold as it is at preſent; which is true, for pro-

* See Howard, 219, 220.

bably its original heat was the ſame as that of many iſlands in the ſame latitude at this day, but ſtill it was too cold for elephants and rhinoceri, and between the climates which they might have then inhabited, and the places they are now found in, too many mountains intercede to ſuppoſe them brought thither by any other means but a *general* inundation. Beſides, Siberia muſt have attained its preſent temperature *at the time* theſe animals were tranſported, elſe they muſt have all long ago putrified.

3dly, Shells known to belong to ſhores under climates very diſtant from each other are in ſundry places found mixed promiſcuouſly with each other; *one ſort* of them, therefore, muſt have been tranſported by an inundation; the promiſcuous mixture can be accounted for on no other ſuppoſition.

Theſe appear to me the moſt unequivocal geologic proofs of a general deluge. To other facts generally adduced to prove it, another origin may be aſcribed; thus the bones of elephants found in Italy, France, Germany, and England, might be the remains of ſome brought to Italy by Pyrrhus or the Carthaginians,

Carthaginians, or of thoſe employed by the Romans themſelves; ſome are ſaid to have been brought to England by Claudius. 4 Phil. Tranſ. Abr. 2d part, 242. When theſe bones, however, are accompanied with marine remains, their origin is no longer ambiguous. Thus alſo the bones and teeth of whales, found near Maeſtricht, are not deciſively of diluvian origin, as whales have often been brought down as low as lat. 48°. 34 Roz. 201. Nay ſometimes they ſtrike on the coaſt of Italy. 1 Targioni Tozzetti 386.

Yet, to explain the leaſt ambiguous of theſe phenomena, without having recourſe to an univerſal deluge, various hypotheſes have been framed.

Some have imagined that the axis of the earth was originally parallel to that of the ecliptic, which would produce a perpetual ſpring in every latitude, and conſequently that elephants might exiſt in all of them. But the ableſt aſtronomers having demonſtrated the impoſſibility of this paralleliſm, it is unneceſſary to examine its conſequences; it only deſerves notice that the obliquity of the equator is rather diminiſhing than increaſing.

creafing. See La Lande in 44 Roz. 212. Befides, why are thefe bones accompanied by marine remains? Others, from the nutation of the earth's axis, have fuppofed that its poles are continually fhifting, and confequently that they might have originally been where the equator now is, and the equator where the poles now are; thus Siberia might have, in its turn, been under the equator. But as the nutation of the earth's axis is retrogreffive every nine years, and never exceeds ten degrees, this hypothefis is equally rejected by aftronomers. 44 Roz. 210. 2 Bergm. Erde Kugel 305. The pyramids of Egypt demonftrate that the poles have remained unaltered thefe three thoufand years.

The 3d hypothefis is that of Mr. Buffon, to which the unfortunate Bailly has done the honour of acceding; according to him the earth, having been originally in a ftate of fufion, and for many years red hot, at laft cooled down to the degree that rendered it habitable. This hypothefis he was led to imagine from the neceffity of admitting that the globe was, at leaft to a certain diftance beneath its furface, originally

nally in a ſoft ſtate; the ſolution of its ſolid parts in water he thought impoſſible, falſely imagining that the whole globe muſt have been in a ſtate of ſolution, whereas the figure of the earth requires the liquidity of it only a few miles beneath its ſurface. Epoques, 10 and 35. If he had trod the path of experiments he would have found both the hardneſs and tranſparency, of what he calls his primitive glaſs, and thinks the primitive ſubſtance of the globe, namely, *quartz*, to be altered in a ſtrong heat with the loſs of 3 per cent. of its weight, and that ſo far from having been a glaſs, that it is abſolutely infuſible. The loſs of weight, he muſt have ſeen, could be aſcribed to nothing elſe but the loſs of its watery particles, and that, therefore, it muſt have been originally formed in water; he would have found that ſome *felſpars* loſe 40 per cent. and others at leaſt 2 per cent. by heat; he would have perceived that *mica*, which he thinks only an exfoliation of quartz, to be in its compoſition eſſentially different. He certainly found their cryſtallization inexplicable, for he does not even attempt to explain it.

But

But waving this, and a multitude of other insuperable difficulties in his hypothesis, and adverting only to the solution he thinks his theory affords, of the phenomenon of the existence of the bones of elephants, and the carcase of a rhinoceros in Siberia, I say it is defective even in that respect. For allowing his supposition that Siberia was at any time of a temperature so suited to the constitution of these animals that they might live in it, yet the remains lately found in that country cannot be supposed to belong to animals that ever lived in it.

1st, Because though they are found at the distance of several hundred miles from the sea, yet they are surrounded by genuine marine vegetables, which shews that they were brought thither together with those vegetables.

2dly, Because they are generally found in accumulated heaps, and it is not to be imagined that while alive they sought a common burial place no more than they at present do in India.

3dly, Because the rhinoceros was found intire and unputrified, whereas if the country

try was warm when he perished this could not have happened.

4thly, Because in no very distant latitude *, namely, that of Greenland, the bones of whales, and not of elephants, are found on the mountains, consequently that latitude must have been in that ancient period sufficiently cold to maintain whales, as it is at this day; and that cold we know to be very considerable, and incompatible with the proximity of a climate suited to elephants. 17 N. Comment. Petropol. 576. 1 Act. Petrop. 55. Renov. 73. Therefore the animals whose remains are now found in Siberia could not have lived in it.

The 4th hypothesis is that of Mr. Edward King, but much amplified and enlarged by Mr. De Luc. This justly celebrated philosopher is of opinion that the actual continents were, before the deluge, the bottom or bed of the ancient ocean, and that the deluge consisted in the submersion of the ancient continents, which consequently form the bottom or bed of our actual oceans, consequently our actual

* The bones were found in latitude 55°.

mountains

mountains were all formed in the antediluvian ocean, and thus ſhells might be left on their higheſt ſummits.

In this hypotheſis the ancient continents muſt have exiſted in thoſe tracts now covered by the Atlantic and Pacific oceans; if ſo, I do not ſee how the elephants could have been brought into Siberia, or a whole rhinoceros found in it: For Siberia being then the bottom of ſome ocean, the ſea muſt have moved *from it* to cover the ſinking continents, inſtead of moving *towards it*, to ſtrew over it their ſpoils.—If it be ſaid that theſe animals were carried into the ſea before the flood, then, aſſuredly, the rhinoceros ſhould have been devoured, and only his bones left.

To ſay nothing of the incompatibility of this ſyſtem with the principal geologic phenomena mentioned in my former Eſſay, and of the deſtruction of at leaſt all the graminivorous fiſh that muſt have followed from their transfer to a ſoil not ſuited to them, it is evidently inconſiſtent with the Moſaic account of this cataſtrophe, which account theſe philoſophers, however, admit.

Moſes aſcribes the deluge to two principal

cipal cauſes, a continual rain for forty days, and the eruption of the waters of the great abyſs. Now to what purpoſe a rain of forty days to overwhelm a continent that was to be immerſed under a whole ocean? He tells us the waters *increaſed* on the continents a certain number of days, *reſted* thereon another period of days, and then *returned*. Do not theſe expreſſions imply a permanent *ground* on which they increaſed and reſted, and from which they afterwards retreated? As the retreat followed the advance, is it not clear that they retreated from the *ſame* ſpaces on which they had before advanced and reſted?

Mr. De Luc replies, that in the 13th verſe of the 6th chapter of Geneſis, it is ſaid the earth ſhould be *deſtroyed*, and that Mr. Michaelis ſo tranſlates it. However it is plain, from what has been juſt mentioned, that Moſes did not underſtand ſuch a deſtruction as ſhould cauſe it to diſappear totally and for ever; he tells us that the waters ſtood 15 cubits over the higheſt mountains; now as he has no where mentioned the antediluvian mountains, but has the poſtdiluvian, it is plain that it is to theſe

his

his narration relates, and these, he tells us, were at the time of the deluge covered with water, and uncovered when the waters diminished; he never distinguished the postdiluvian from the antediluvian, and therefore must have considered them as the same.

Nor did Noah himself believe the ancient continents destroyed, for he took the appearance of an olive branch to be a sign of the diminution of the flood. This he certainly believed to have grown on the ancient continent, and could not expect it to have shot up from the bottom of the sea.— M. De Luc tells us that this olive grew on an antediluvian island, and that these islands, being part of the antediluvian ocean, were not flooded—it is plain, however, Noah did not think so, else he would not judge the appearance of the olive to be a sign of the diminution of the waters.— Where is it mentioned, or what renders it necessary to infer that islands existed before the flood? If islands did exist, and were to escape the flood, so might their inhabitants also, contrary to the express words of the text.

It

It would ſurely be much more convenient for Noah, his family, and animals, to have taken refuge in one of them, than to remain pent up in the ark.

The dove, Moſes tells us, returned the firſt time ſhe was let out of the ark, *finding no place whereon to reſt her foot*; ſhe conſequently could not diſcover the iſland, whereas the raven never returned, plainly becauſe he found carcaſſes whereon to feed, therefore theſe carcaſſes were not ſwallowed up, as Mr. De Luc would have it. Moſes tells us that at the ceſſation of the flood the fountains of the deep were ſtopped or ſhut up; therefore, in his apprehenſion, inſtead of the ancient continents ſinking into the deep, the waters of the abyſs flowed from their ſources upon that continent and again returned; from all which it follows that this hypotheſis is as indefenſible as the foregoing.

Paſſing over the ſyſtems of Burnet, Woodward, and Whiſton, which have been repeatedly refuted, I recur to the account of this great revolution given by Moſes himſelf, taken in its plain literal ſenſe, as the only one that appears perfectly conſiſtent,

ſiſtent, with all the phenomena now known, of which I ſhall find occaſion to mention many; he plainly aſcribes it to a ſupernatural cauſe, namely the expreſs intention of God to puniſh mankind for their crimes. We muſt therefore conſider the deluge as a miraculous effuſion of water, both from the clouds and from the great abyſs; if the waters, ſituated partly within and partly without the caverns of the globe, were *once* ſufficient to cover even the higheſt mountains, as I have ſhewn in the former Eſſay, they muſt have been ſufficient to do ſo a ſecond time when miraculouſly educed out of thoſe receptacles.

Early geologiſts, not attending to theſe facts, thought all the waters of the ocean inſufficient; it was ſuppoſed that its mean depth did not exceed a quarter of a mile, and that only half of the ſurface of the globe was covered by it; on theſe *data* Keil computed that twenty-eight oceans would be requiſite to cover the whole earth to the height of four miles, which he judged to be that of the higheſt mountains, a quantity at that time conſidered as extravagant and incredible, but a further progreſs in mathematical

mathematical and phyſical knowledge has ſince ſhewn the different ſeas and oceans to contain at leaſt forty-eight times more water than they were ſuppoſed to do.

Mr. De La Place, calculating their average depth, not from a few vague and partial ſoundings, for ſuch they have ever been (the polar regions having been never ſounded, particularly the Antarctic), but from a ſtrict application of the theory of tides to the height to which they are known to riſe in the main ocean, demonſtrates that a depth, reaching only to half a league, or even two or three leagues, is incompatible with the Newtonian theory, as no depth under four leagues can reconcile it with the phenomena *.—The vindication of the Moſaic hiſtory does not require near ſo much. The extent of the ſea is known to be far greater than Keil ſuppoſed, that of the earth ſcarcely paſſing $\frac{1}{3}$ of the ſurface of the globe. 1 Lulolff, p. 233, § 244.

The poſſibility and reality of the deluge being thus eſtabliſhed, I ſhall next endeavour to trace its origin, progreſs, and ſtill

* Mem. Paris, 1776, p. 213.

permanent consequences. That it originated in and proceeded from the great southern ocean below the equator, and thence rushed on the northern hemisphere, I take to be a natural inference from the following facts:

1st, The southern ocean is the greatest collection of waters on the face of the globe *.

2d, In the northern latitudes beyond 45° and 55° we find the animal spoils of the southern countries, and the marine exuviæ of the southern seas, but in the southern latitudes we find no remains of animals, vegetables or shells belonging to the northern seas, but those only that belong to the neighbouring seas. Thus in Siberia, to return to the already frequently mentioned phenomenon, we find the remains of elephants and rhinoceri accompanied by marine vegetables, and also with shells that do not belong to the northern ocean. 1 Epoques, 418. 1 Act. Petropol. 55. They must therefore have been conveyed thither

* In many parts of the south sea no bottom can be found. Mem. Philosoph. &c. concernant la decouverte de l'Amerique. Paris, 1787.

by the more diſtant Indian ſea overflowing theſe parts; as the elephants very naturally crowded together on the approach of the inundation, they were conveyed in flocks, and hence their bones are found in accumulated heaps, as ſhould be expected*. But in Greenland, which is ſtill more diſtant, only the remains of whales are found on the mountains. Crantz Hiſtoire Generale des Voy. vol. xix. 105. So in the ſouthern latitudes, as at Talcaguana in Chili, latitude 36° S. the ſhells found on the tops of the hills are thoſe of the neighbouring ſea. 2 Ulloa Voy. p. 197. So thoſe found on the hills between Suez and Cairo are the ſame as thoſe now found in the Red Sea. Shaw's Voyages, vol. ii. and thoſe that exiſt in the calcareous rocks that ſerve as a baſis to Egypt, are ſuch as are found in the neighbouring ſeas. Volney *Egypt*, 8, 9.

3dly, The traces of a violent ſhock or impreſſion from the ſouth are as yet perceptible in many countries. This Mr. Pa-

* The elephantine remains found in Ruſſia were conveyed thither from Perſia.

trin attests as to the mountains of Dauria on the south east limits of Siberia; he tells us that the more eastern extremities of the mountains appear to be broken off by the impetuosity of an ancient ocean rushing from east to west; that the fragments carried to the west in some measure protected the more western. 38 Roz. 230. 238. And that in general the mountains of this country were so disordered (by the shock) that the miners are obliged to work at hazard. Ibid. 226. Steller makes the same remarks on the mountains of Kamschatka, 51 Phil. Transf. Part ii. p. 479. and observed that the S. end of Bering's island is much more shattered than the N. extremity. 2 Nov. Nord. Beytr. 262. Storr, Hœpsner, and Saussure, inform us that the inundation that invaded Swisserland proceeded from the south, but its impression was modified by another event which I shall presently mention. 1 Helvet. Mag. 173. 175. and 293. 4 Helvet. Magaz. 307. Lasius tells us that the mountains of the Hartz suggest the same inference. Hartz. 95. Many of the fish found on the slates of La Bolca

Bolca are thoſe of the South Sea. 5 Ir. Accad. 288.

4thly. The very ſhape of the continents which are all ſharpened towards the ſouth, where waſhed by the ſouthern ocean, indicate that ſo forcible an impreſſion was made on them as nothing but the mountains could reſiſt, as the Cape of Good Hope, Cape Cormorin, the ſouthern extremity of New Holland, and that of Patagonia; Foſter's Obſervations, p. 11, 12.

To theſe geological proofs perhaps I may be permitted to add the tradition of the orthodox Hindus, that the globe was divided into two hemiſpheres, and that the ſouthern was the habitation of dæmons that warred upon the gods. 3 Aſiatic Reſearches, 51 and 52. This war is commonly thought to be an allegorical deſcription of the *flood*, and hence the olive branch, denoting a diminution of the flood, became a ſymbol of peace.

Did not Noah reſide on the borders of the ſouthern ocean, otherwiſe he could not ſee that the great abyſs was opened? and did not an inundation from the ſouth-eaſt drive the ark north-weſt to the mountains of Armenia? These conjectures are at least consistent with the most probable notions of the primitive habitation of man, which I take to be near the sources of the Ganges, (as Josephus expressly mentions) the Bourampooter and the Indus, from which, as the temperature grew colder, mankind descended to the plains of India.

This unparalleled revolution, Moses informs us, was introduced by a continual rain for forty days. By this the surface of the earth must have been loosened to a considerable depth; its effects may even have been in many instances destructive; thus in August, 1740, several eminences were swept away, nay the whole mountain of Lidsheare, in the province of Wermeland in Sweden, was rent asunder by a heavy fall of rain for only one night. 27 Schwed. Abhand, 93. This loosening and opening of the earth was in many places, where the marine inundation stagnated, an useful operation to the soil subsequently to be formed, as by these means shells and other marine exuviæ were introduced into it which rendered it more fertile. By this rain also, the salt water was diluted, and its perni-

cious

cious effects both to soil and fresh water fish in great measure prevented. The destruction of animals served the same purposes, and might, in many instances, be necessary to fertilize a soil produced by the decomposition of primary mountains; from the animals thus destroyed the phosphoric acid found in many ores may have originated.

But the completion of this catastrophe was undoubtedly effected, as Moses also states, by the invasion of the waters of the great abyss, most probably, as I have said, that immense tract of ocean stretching from the Philipine islands, or rather from the Indian continent on the one side to terra firma on the other, and thence to the southern pole; and again from Buenos Ayres to New Holland, and thence to the pole.—Tracing its course on the eastern part of the globe, we shall see it impelled northwards with resistless impetuosity against the continent which at that time probably united Asia and America. 1 Physical. Arbeit. p. 13. 1 Nev. Nord. Beytr. 278 and 294. This appears to have been torn up and swept away (except the islands that

that ſtill remain, ſome of which are primitive mountains, and ſeem torn from the continent. Staunton's Voy. to China, 4to. p. 310.) as far north as latitude 40°; its further progreſs appears to have been ſomewhat checked by the lofty mountains of China and Tartary, and thoſe on the oppoſite American coaſt; here then it began to dilate itſelf over the collateral countries; the part checked by the Tartarian mountains forming, by ſweeping away the ſoil, the deſart of Coby, while the interior or middle torrent preſſed forward to the pole; but the interior ſurge being ſtill more reſtricted by the contiguous, numerous, and elevated mountains of eaſtern Siberia and America, muſt at laſt have ariſen to a height and preſſure which overbore all reſiſtance, daſhing to pieces the heads of thoſe mountains, as Patrin and Steller remark, and bearing over them the vegetable and animal ſpoils of the more ſouthern, ravaged or torn up continents, to the far extended and inclined plains of weſtern Siberia, where its free expanſion allowed it to depoſit them. Hence the origin of the bones and tuſks of elephants and rhinoceri found in the plains, or inconſiderable ſandy or marly eminences in the north weſtern parts of Siberia, as Mr. Pallas rightly judges.

If now, returning to the ſouth, we contemplate the effects of this overwhelming invaſion on the more ſouthern regions of India and Arabia, we ſhall, where the coaſts were undefended by mountains, diſcover it excavating the gulphs of Nanquin, Tonquin, and Siam, the vaſt Bay of Bengal, and the Arabic and Red Seas. That the ſouthern capes, promontories, and headlands, were extenuated to their preſent ſhape by the deluge, and not by tides or the currents ſtill obſerved in thoſe ſeas, may be inferred from the inefficacy of thoſe feebler powers to produce any change in them for many paſt centuries.

The chief force of the inundation ſeems to have been directed northwards in the meridians of from 110 to 200 eaſt of London. In the more weſtern tracts it appears to have been weaker; the plains of India I ſuſpect to have been leſs ravaged, or perhaps their ſubſequent fertility may have been occaſioned by the many rivers by which that happy country is watered. Not

ſo

ſo thoſe of Arabia; their ſolid baſis, reſiſting the inundation, was obliged to yield its looſer ſurface, and remains even now a ſandy deſart, while the interior more mountainous tracts, intercepting, and thus collecting, the waſhed off ſoil, are, to this day, celebrated for their fertility, 2 Niebuhr, 45 and 320. Iriſh edition. To a ſimilar tranſportation of the ancient vegetable ſoil, the vaſt ſandy deſarts of Africa, and the barrenneſs of moſt of the plains of Perſia, may be attributed.

The progreſs of the Siberian inundation once more claims our attention; that it muſt have been here for ſome time ſtationary may be inferred from its confinement between the Altaiſchan elevation on the ſouth, and the Ouralian mountains on the weſt, and the circumpolar mountains on the ſide of Greenland. Hence the excavations obſerved on the northern parts of the former, and the abrupt declivities on the eaſtern flanks of the latter, while the weſtern diſcover none. New reinforcements from the ſouth-eaſt muſt at length have ſurmounted all obſtacles; but the ſubſequent ſurges could not have conveyed ſuch

ſuch a quantity of ſhells or marine productions as the firſt, and hence, though many are found on the more northern *plains*, ſcarce any are found on the great Altaiſchan elevation.

The maſs of waters now collected and ſpread over the Arctic regions, muſt have deſcended partly ſouthwards over the deſarts of Tartary, into countries with which we are too little acquainted to trace its ravages, but from the oppoſition it muſt have met in theſe mountainous tracts, and the repercuſſion of their craggy ſides, eddies muſt have been formed to which the Caſpian, Euxine, and other lakes may have owed their origin. Part alſo muſt have extended itſelf over the vaſt tracts weſt of the Ourals, and there expanded more freely over the plains of Ruſſia and Poland down to latitude 52°, where it muſt have met with and be oppoſed by the inundation originating in the weſtern parts of the Pacific ocean, this ſide the Cape of Good Hope, and thence impelled northwards and weſtwards in the ſame manner as the eaſtern inundation already deſcribed, but with much leſs force, and ſweeping the conti-

nents

nents of South America (if then emerged) and of Africa, conveying to Spain, Italy, and France, and perhaps ſtill farther north, elephants and other animals, and vegetables hitherto ſuppoſed partly of Indian and partly of American origin.

That the courſe here aſſigned is not imaginary, appears from the ſhells, vegetables, and animal remains of thoſe remote climates, ſtill found in Europe, and from the diſcovery both of the European and the American promiſcuouſly mixed with each other at Fez. 1 Bergman Erde Kugel, 252. 249.

So in Germany, Flanders, and England, the ſpoils of the northern climates, and thoſe of the ſouthern alſo, are equally found; thus the teeth of arctic bears, and bones of whales, as well as thoſe of animals of more ſouthern origin, have been diſcovered in thoſe parts. Sauſſure alſo has found traces in Swiſſerland of an inundation from the north. 7 Sauſſ. 216.

The effect of the encounter of ſuch enormous maſſes of water, ruſhing in oppoſite directions, muſt have been ſtupendous, it was ſuch as appears to have ſhaken and ſhattered

ſhattered ſome of the ſolid vaults that ſupported the ſubjacent ſtrata of the globe. To this concuſſion I aſcribe the formation of the bed of the Atlantic from latitude 20° ſouth up to the north pole. The bare inſpection of a map is ſufficient to ſhew that this vaſt ſpace was hollowed by the impreſſion of water; the protuberance from Cape Frio to the river of the Amazons, or La Plata in South America, correſponding with the incavation on the African ſide from the river of Congo to Cape Palmas; and the African protuberance from the Straits of Gibraltar to Cape Palmas, anſwering to the immenſe cavity between New York and Cape St. Roque. The depreſſion of ſuch a vaſt tract of land cannot appear improbable when we conſider the ſhock it muſt have received, and the enormous load with which it was charged. Nor is ſuch depreſſion and abſorption unexampled, ſince we had frequent inſtances of mountains ſwallowed up, and ſome very lately in Calabria. *

The

* The Bay of Galway appears to have been originally a granitic mountain, ſhattered and ſwallowed during this cataſtrophe, for fragments of granite are found

The wreck of ſo conſiderable an integrant part of the globe muſt of neceſſity have convulſed the adjacent ſtill ſubſiſting continents previouſly connected with it, rent their ſtony ſtrata, burſt the ſtill more ſolid maſſes of their mountains, and thus in ſome caſes formed, and in others prepared, the inſular ſtate to which theſe fractured tracts were reduced; to this event, therefore, I think, may be aſcribed the bold, ſteep, and abrupt weſtern coaſts of Ireland, Scotland, and Norway, and the numerous iſles that border them, as well as many of thoſe of the Weſt Indies. 2 Wms. 162. The Britannic iſlands ſeem to have acquired their inſular ſtate at a later period, though it was probably prepared by this event;

found on its northern ſhore, though none in the neighbouring mountains, which are chiefly argillitic. Alſo a vaſt maſs of granite called the Gregory, lately ſtood on one of the iſles of Arran, 100 feet at leaſt above the level of the ſea, 10 or 12 feet high, as many broad, and about twenty in length. Though the whole maſs of the iſland conſiſts of compact limeſtone, and no granitic hill within 8 or 10 miles of it. This was ſhattered by lightning in 1774.

The general motion of the Atlantic is as yet from N. to S. Howard, 355. Staunton's Voyage to China.

but

but the baſaltic maſſes on the Scotch and Iriſh coaſts, and thoſe of Feroe, appear to me to have been rent into pillars by this concuſſion.

During this elemental conflict, and the craſh and ruin of the ſubmerged continent, many of its component parts muſt have been reduced to atoms, and diſperſed through the ſwelling waves that uſurped its place. The more liquid bitumens muſt by the agitation have intimately mixed with them. They muſt alſo have abſorbed the fixed air contained in the bowels of the ſunk continent; and further, by this vaſt continental depreſſion, whoſe derelinquiſhed ſpace was occupied by water, the level of the whole diluvial ocean muſt have been ſunk, and the ſummits of the higheſt mountains muſt then have emerged. In this ſtate of things it is natural to ſuppoſe that if iron abounded in the ſubmerged continent, as it does at this day in the northern countries of Sweden, Norway, and Lapland adjacent to it, its particles may have been kept in ſolution by the fixed air, and the argillaceous, ſiliceous, and carbonaceous particles may have been

 long

long ſuſpended. Theſe muddy waters mixing with thoſe impregnated with bitumen, the following combinations muſt have taken place: 1°. If carbonic matter was alſo contained in the water, this uniting to the bitumen, muſt have run into maſſes no longer ſuſpenſible in water, and formed ſtrata of coal. 2do. The calces of iron by the contact of bitumen were in great meaſure gradually reduced, and together with the argillaceous and ſiliceous precipitated on the ſummits of ſeveral of the mountains not yet emerged, and thus formed baſaltic maſſes, that, during deſiccation, ſplit into columns; in other places they covered the carbonaceous maſſes already depoſited, and by abſorbing much of their bitumen rendered them leſs inflammable, and hence the connexion which the ſagacious Werner obſerved between baſalts and coal. The fixed or oxygen air, erupting from many of them, formed thoſe cavities, which being filled by the ſubſequent infiltration of ſuch of their ingredients as were ſuperfluous to their baſaltic ſtate, formed calcedonies, zeolytes, olivins, baſaltines, ſpars, &c. Hence moſt of the mountains of Sweden

that

that afford iron, afford alſo bitumen *. Hence alſo the aſphalt found with Trap † and under baſalts ‡, and in balls of calcedony found in Trap §.

This I take to be the *laſt* ſcene of this dreadful cataſtrophe, and hence ſhells are ſeldom found in theſe baſalts, they having been previouſly depoſited, though ſome other lighter marine vegetable remains have ſometimes been found in them ||; ſome argillaceous or ſandſtone ſtrata may alſo have been depoſited at this period.

On this account, however, of the formation of the baſalts which crown the ſummits of ſeveral lofty peaks, I lay no more ſtreſs than it can juſtly bear; I deliver it barely as an hypotheſis more plauſible than many others.

It has been objected to the Moſaic account that the countries near Ararat are too cold to bear olive trees. Tournfort, who firſt made this objection, ſhould recollect, that at this early period, the Caſpian

* Berg. Jour. 1789, p. 2005.

† Berolding on Mercury, p. 38 and 240.

‡ Von. Salis 171. 3. Noſe. 146.

§ 10 Naturforſch. 43.

|| I have lately ſeen ſhells in the baſalts of Ballycaſtle.

pian and Euxine ſeas were joined, as he himſelf has well proved. This circumſtance ſurely fitted a country lying in the 38th degree of latitude to produce olives (which now grow in much higher latitudes) at preſent chilled only by its diſtance from the ſea.

A more plauſible objection ariſes from the difficulty of collecting and feeding all the various ſpecies of animals now known, ſome of which can exiſt only in the hotteſt, and others only in the coldeſt climates; it does not however appear to me neceſſary to ſuppoſe that any others were collected in the ark but thoſe moſt neceſſary for the uſe of man, and thoſe only of the graminivorous or granivorous claſſes, the others were moſt probably of ſubſequent creation. The univerſality of the expreſſions, Gen. chap. vi. ver. 19. "Of every living thing, "of all fleſh, two of every ſort ſhalt thou "bring into the ark," ſeem to me to imply no more than the ſame general expreſſions do in Gen. chap. i. ver. 30. "And "to *every* beaſt of the earth, and to *every* "fowl of the air, have I given *every* green "herb for meat;" where it is certain that

only

only graminivorous animals are meant. At this early period ravenous animals were not only not neceſſary but would have been even deſtructive to thoſe who had juſt obtained exiſtence, and probably not in great numbers. They only became neceſſary when the graminivorous had multiplied to ſo great a degree that their carcaſſes would have ſpread infection. Hence they appear to me to have been of poſterior creation; and to this alſo I attribute the exiſtence of thoſe that are peculiar to America and the torrid and frigid zones.

The atmoſphere itſelf muſt have been exceedingly altered by the conſequences of the flood. Soon after the creation of vegetables, and in proportion as they grew and multiplied, vaſt quantities of oxygen muſt have been thrown off by them into the then exiſting atmoſphere without any proportional counteracting diminution from the reſpiration or putrefaction of animals, as theſe were created only in pairs, and multiplied more ſlowly; hence it muſt have been much purer than at preſent; and to this circumſtance, perhaps, the longevity of the antediluvians may in great mea-

 ſure

ſure be attributed. After the flood the ſtate of things was perfectly reverſed, the ſurface of the earth was covered with dead and putrifying land animals and fiſh, which copiouſly abſorbed the oxygenous part of the atmoſphere and ſupplied only mephitic and fixed air; thus the atmoſphere was probably brought to its actual ſtate, containing little more than one-fourth of pure air and nearly three-fourths of mephitic. Hence the conſtitution of men muſt have been weakened and the lives of their enfeebled poſterity gradually reduced to their preſent ſtandard. To avoid theſe exhalations it is probable that the human race continued for a long time to inhabit the more elevated mountainous tracts. Domeſtic diſturbances in Noah's family, briefly mentioned in holy writ, probably induced him to move with ſuch of his deſcendants as were moſt attached to him, to the regions he inhabited before the flood, in the vicinity of China, and hence the early origin of the Chineſe monarchy.

ESSAY III.

OF SUBSEQUENT CATASTROPHES.

A SHOCK ſo violent and univerſal as that which pervaded the globe during the diluvian revolution muſt have produced not only innumerable alterations in the whole extent of its ſurface, endleſs to detail (and belonging rather to the natural hiſtory of its particular geographic diviſions than to a general ſurvey of the whole), but alſo have prepared, by looſening its baſis, many other changes that took effect ſome centuries after, as I have already hinted; of theſe, however, ſome are ſo important, by their connexion either with the paſt tranſactions of the inhabitants of the globe, or with the actual external appearance or ſubterraneous ſtate of the countries with which we are beſt acquainted, that they cannot be totally paſſed over in ſilence; ſuch are the total ſeparation of Aſia from America, the coarctation of the Baltic, the ſeparation of the

the Caſpian from the Black Sea, and the junction of this with the Mediterranean, and of the Mediterranean with the Ocean; and, laſtly, the ſeparation of Ireland from Britain, and of Britain from the Continent.

Theſe events are either totally omitted by hiſtorians, or only ſlightly mentioned by that hiſtory, from its mixed nature, ſtiled *fabulous*; in ſuch caſes imagination has commonly taken up the pen of the hiſtorian, but poſſeſſing neither the inclination nor the talents neceſſary for converting natural hiſtory into *romance*, the account I ſhall give of them muſt be very ſhort, and ſuch as is ſuggeſted by the moſt probable traditions or actual appearances.

1°. Of the ſeparation of Aſia from America, we have no traditional account; it is certain, however, that they were once joined, as the inhabitants of the former paſſed into the latter; they ſtill approach each other very nearly in the northern parts, and the intermediate ſpace is filled with iſlands. The junction immediately after the flood I ſtated in my former Eſſay to have reached ſo low as latitude 40°, becauſe I find America peopled by nations of different

different characters, degrees of civilization, and languages; such varieties could scarcely occur among a people inhabiting the same climate, and formed to the same habits. As Cain was originally banished from his brethren, so I imagine malefactors anciently were into the colder deserted climates, by their civilized brethren in Asia, and thus the Esquimaux originated, that are equally found in both continents, and other savages; and hence the predatory disposition of the Scythian tribes. That the postdiluvian junction did not reach lower than latitude 40°, I infer from the absence of elephants from the warmer regions of America.

The utter separation of both continents was most probably the effect of excavations by volcanoes; at least this cause is adequate to such an effect, and it still exists in the most northern parts. The superior fertility of the western coast of America may arise from the lavas ejected on that coast.

2d. That the Baltic in all its branches was anciently much more extended than at present, many reasons induce us to believe; but principally the state in which we at

present

present find the immense plains of Southern Russia from Petersburgh to Pultowa.

These plains for some hundred miles to the south of Petersburgh are still a morass, and farther southwards they are covered with sand, pebbles, and petrified shells *. This water is not, indeed, salt, but neither was the Baltic so originally, and is but slightly so at present, for it seems to have been formed by the confluence of the various rivers that flow into it, which at last burst a passage into the German sea; by communication with this it became salt. At present there are three passages by which they communicate, at first probably but one; to the opening of the two last the reduction of this sea to its present limits is owing.

3d. The ancient communication between the Caspian, the Lake of Aral, and the Black Sea, before the opening of the Thracian Bosphorus, which enabled these seas to discharge themselves into the Mediterranean, is rendered highly probable, if not

* Schwed. Abhand. 1773, 181, &c. 2 Bergm. Erde kugel. 215.

demonstrated,

demonſtrated, by Pallas, 3 Pallas Reiſe, 368. He obſerved, that between the rivers Sarpa and Volga, from Zarzycin down to the Caſpian Sea, the land ſlopes with conſiderable indentures and abrupt promontories as if it had been an ancient coaſt, and continues on the ſame level on the eaſt of the Wolga in the ſandy deſert of Narym, and in the more ſouthern ſteppes or deſerts between the Wolga and the Jaick or Ural. The ſhells which abound in this extenſive *flat* exactly reſemble thoſe of the Caſpian, and are different from thoſe of the adjacent rivers; theſe deſerts are moreover covered with ſand, contain abundance of ſalt and ſalt lakes, and produce only ſuch vegetables as grow on ſalt marſhes; whereas the upland that borders this flat contains a genuine black fertile ſoil, and no ſhells reſembling thoſe of the Caſpian, but only thoſe of ancient date, ſuch as thoſe found in other countries, and of the ſorts called *Pelagicæ* that belong to the deepeſt ſeas. Hence he collects that the level of the Caſpian before it was reduced to its preſent limits was ninety feet higher than at preſent; thus it was enabled to communicate

communicate with the Euxine by the ſea of Aſoph, In this ſtate it muſt have remained from the period of the deluge until about 1800 years before our Æra, the moſt probable date of the ſeparation of theſe ſeas, as Mr. Foſter has ſhewn in a learned memoir in the Gottingen Magaz. 1780. p. 140.

Buffon pretends the rupture of the iſthmus that ſeparated this ſea from the Mediterranean preceded the deluge mentioned by Moſes, merely it ſhould ſeem to contradict that moſt authentic hiſtorian, for he aſſigns no reaſon independent of his own hypotheſis, and is deſtitute of the ſupport even of the moſt fabulous tradition. 1 Epoques, 8vo. p. 291.

The rupture of the iſthmus was probably ſudden and total, and conſequently effected by an earthquake. To diſcover its effects, we muſt firſt conſider the antecedent ſtate of the Mediterranean:

The Mediterranean, before its union with the Black Sea and the Ocean, was moſt probably a baſon much narrower and ſhallower than at preſent; for though it received ſeveral conſiderable rivers, the Nile, the

 Rhone,

Rhone, and the Po *, yet ſince even now evaporation from its ſurface is ſufficient to prevent it from overflowing, notwithſtanding that the Ocean on the one ſide, and the Euxine on the other, flow into it, we may well ſuppoſe that when it communicated with neither, evaporation kept its level much lower; when, therefore, by the rupture of the Thracian iſthmus on the one ſide, and of the African which joined Ceuta with Gibraltar on the other, the waters of both were poured in upon it, an immenſe preſſure took place on its bed, under which it ſunk and fell into the inferior cavity of the globe; during this tremendous tumult the iſlands of Sicily, Sardinia, Corſica, and thoſe of the Archipelago were torn off, and Italy was lengthened to its preſent ſhape. Fortis *Dalmatia*, 173. The neighbouring ſhores of France and Spain, and more eſpecially thoſe of Africa as being much lower, and thoſe of Greece and Aſia, muſt have been inundated to a great extent, and hence the ſaline ſub-

* Allowance ſhould be made for rains as well as for rivers. See Howard's Theory of the Earth, 34.

ſtances

ſtances ſtill exiſting in the adjacent parts of Africa, &c.

As the ſouthern parts of Italy ſtill abound in ſulphur and other inflammable ſubſtances, ſo probably did the contiguous parts of the bed of the Mediterranean, and by the immenſe friction which they muſt have ſuffered during this fall and the hollows that interceded the abrupted maſſes the firſt ſubterraneous fires might have been kindled and the beds of the actual volcanoes prepared, which however did not probably acquire ſufficient ſtrength to burſt through the incumbent earthy ſtrata until ſome ages after, as I conjecture from the ſilence of Homer with reſpect to Ætna, whoſe wonders, had they exiſted in his time, he probably would not have overlooked.

The ſeparation of Sicily from Italy is vouched by ancient traditions, as may be ſeen in Pliny, Ovid, and Claudian.

Zancle quoque juncta fuiſſe
Dicitur Italiæ, donec confinia pontus
Abſtulit, & media tellurem reppulit unda.

Ovid. Metam. Lib. 15. *v.* 290.

Trinacria

Trinacria quondam
Italiæ pars una fuit, ſed pontus & æſtus
Mutavere ſitum, rupit confinia Nereus
Victor, & abſciſſo interluit æquore montes. *Claudian. De Rapt. Proſerp. Lib.* 1.

The ſteep abrupt coaſts from Genoa to Leghorn, deſcribed by Ferber in his twenty-ſecond letter, muſt be aſcribed to the rupture of the ſtrata, as tides, ſcarcely ſenſible in this ſea, cannot be even ſuſpected of having acted ſo powerfully upon them. The rapidity of the Rhone, and of moſt of the rivers that fall into this ſea on the European ſide, alſo indicate the great inclination of the ſtrata of the interior countries towards it, a natural conſequence of the depreſſion of their primitive ſupport. The mountains of Swiſſerland diſcover alſo veſtiges of a ſhock on the ſouth eaſt, as I have already noticed, the detail of which I leave to the many excellent geologiſts of that country.

The communication of the Euxine with the Ocean by means of the Mediterranean, being thus formed, its level gradually ſubſided, the canal which joined it with the

Caſpian dried up; as few great rivers fall into this (only the Wolga and the Ural) it was ſoon reduced by evaporation to its preſent level, which is ſaid to be lower than that of any other ſea, and thus the ſalt deſerts that border it, were formed, and its ſeparation from the Aral effected.

5°. The entire ſeparation of Great Britain from the Continent muſt have happened long after the deluge, and that of Ireland from Great Britain at a ſtill later period; for wolves and bears were anciently found in both, and theſe muſt have paſſed from the Continent into Britain, and from this into Ireland, as their importation cannot be ſuſpected. Theſe events, as I already ſaid, muſt have been prepared and have commenced by the ſhock communicated during the rupture and depreſſion of the bed of the Atlantic. The divulſive force that ſeparated Britain from Germany, ſeems to have been directed from north to ſouth, but gradually weakened in its progreſs. Hence that iſland is ſharpened to the northwards, but the impreſſion muſt have been conſiderably weakened by the oppoſition of the granitic mountains that form the Shetland

Shetland and Orkney Isles. The looser structure of the calcareous or argillaceous and arenaceous materials of the more southern parts offered less resistance, was more easily preyed upon, and gave way to, what is now called, the German ocean, while these materials themselves were spread over Westphalia, &c. or formed the subsoil of Flanders, Holland, and the sand banks on its coast. The rupture of the isthmus that joined Calais and Dover was probably effected by an earthquake at a later period, and gradually widened by tides and currents. Ireland was protected by Scotland from the violence of the northern shock, hence its separation from Scotland appears to have been late and gradual. That from England was probably diluvial and effected by a southern shock.

All these changes happened at least three thousand six hundred years ago, and I see no reason to think that the general level of the ocean has since been altered, but that of the continents seems to have varied considerably, being in some places higher and in others lower than anciently.

The depression of continents originates

 from

from two caufes; the firft is the diminution of the waters that anciently pervaded them; the fecond is the fliding away of the inferior argillaceous ftrata; to fay nothing of caufes merely contingent, as earthquakes and inundations.

After the deluge, the earth on which it refted 150 days, and from which it very flowly retired, muft have been drenched with water; vegetation quickly enfued, and twenty centuries ago moft countries are known to have been covered with trees, and many until a much later period; in thefe circumftances an infinity of fmall rivulets muft have been formed, which poured their ftreams into the great rivers arifing from their confluence; while the furface of the earth was protected by forefts from the immediate influence of the fun, the moifture replenifhing its interior muft have remained. Again, as we fee the craggy fummits of many of the higheft mountains, now decompofing, being corroded by air and moifture, we muft fuppofe that the fame caufes have operated in the moft ancient times, and that previous to their action thefe fummits were much

 higher,

higher, and conſequently better fitted for collecting vapours than at preſent, but in modern times, from the extenſion of agriculture to countries where it never before was introduced, and the increaſing multiplication of mankind, the foreſts have in great meaſure diſappeared; the earth laid bare yields its waters to evaporation; the mountains lowered, no longer collect or condenſe the ſame quantity of vapour, the rivulets ceaſe, or are reduced to rills, and the earth freed from its pervading moiſture has naturally ſunk to a lower level. Hence moſt rivers were anciently much more conſiderable than at preſent, as we may ſee by the wide extended vales through which they run, the monuments of their ancient magnitude. Not only bogs have in ſome inſtances been known to ſlide over lower adjacent plains, but even portions of land ſeemingly more ſolid. Thus in France, near Meudon, in 1787, a whole ſide of a hill, covered with trees, deſcended 50 feet upon, and covered to the height of 70 feet, a neighbouring plain; its deſcent laſted ſix years. 43 Roz. 19. It was occaſioned by laying bare a bed of clay, which, by imbibing rain was tumefied and loosened. In Bohemia also, so late as the year 1770, a great part of the mountain of Ziegenberg slid down 38 fathom to the Elbe with its trees still standing partly erect and partly inclined. This was ascribed to the solution of its inferior stratum, an argillaceous sand-stone, whose argillaceous part was carried off by rain to which it had incautiously been exposed. Reuss *Bohemia*, 55 *.

In some places the surface of the earth has been elevated by the particles carried down by rain from greater elevations, or the gradual disintegration of the stony substances that covered them, or by various local and contingent events. Hence many remains of antiquity deposited on the ancient surface are now found at considerable depths, particularly in countries long devasted.

The effects of volcanoes in altering the face of the globe seem to me much more circumscribed than many late writers have asserted: few mountains, and those easily

* See also Mem. Par. 1746, 1119, in 8vo. in notes; and Mem. Par. 1769, p. 500, in 8vo. and 5 Nev. Nord. Beytr. 260, for that of Tauria in 1786.

distinguish-

diſtinguiſhable, owe their origin to them; neither Veſuvius nor Ætna were formed by them, as is evident by the maſs of Neptunian ſtones that compoſe them; their effuſions ſeldom reach to a great diſtance, none above 100 miles. Of ancient volcanoes now extinct, few can be traced by undoubted characters or hiſtoric accounts. Sidonius Apollinaris, lib. 7. Epiſt. 1. mentions one near Vienne in Dauphiné which burned about the year 469; but by his account it is not clear whether it was a volcano or a pſeudo volcano; I ſuſpect the latter.

Tacitus alſo, at the end of the 13th Book of his Annals, mentions a volcano, or rather pſeudo volcano, that ravaged the country of the Jutiones; this is ſaid by many to be that of the Ubians; if ſo, it muſt have been in the neighbourhood of Cologn, where ſome pſeudo volcanic remains are traced.

On no ſubject have philoſophers been leſs cautious of the deluſions of imagination than on this of volcanoes: the aſtoniſhment excited by their awful phenomena ſeems to have affected even the underſtanding of ſome of their ſpectators. The author of many celebrated treatiſes on volcanoes, lately travelling into Scotland, exclaimed, at the ſight of every black ſtone he met with, that it was *lava*, as I was informed by one of his companions *; even the very excellent Sir William Hamilton has frequently been ſeduced from the ſimple path of obſervation, to which, notwithſtanding, he profeſſed to adhere, into the mazes and errors of a baſeleſs ſyſtem †. In a letter to Sir John Pringle, May 1776, he tells us that " Wherever baſaltic pillars like thoſe " of the Giant's Cauſeway in Ireland are " found, there without doubt a volcano " muſt have exiſted, for they are mere lava." At preſent, however, I believe none will pretend that the volcanic origin of theſe pillars is out of the reach of doubt. He tells us that Veſuvius and Ætna were formed by a ſeries of volcanic eruptions ‡, though there is no certain proof that the former was ſo formed, and it is demonſtra-

* Mr. Macie, a gentleman of the moſt exact and extenſive mineralogical knowledge.

† Ouvres de Hamilton, p. 10.

‡ Ouvres de Hamilton, p. 11.

ble

ble that the latter exiſted as a mountain before it became a volcano. Padre Torre, who has given a good deſcription of Veſuvius, inſiſts that its primitive ſtamina, if I may ſo call them, are not volcanic, but that it ſhould rather be conſidered as an extenſion of the Appenines; the number of Neptunian ſtones it throws up, as may be ſeen in Gioeni's Lithography of Veſuvius, confirms this opinion; that the calcareous ſtrata are covered to a great depth with lava cannot be doubted, but that the whole maſs of Veſuvius conſiſts of volcanic ejections has not been proved; it is ſaid that in ſinking a well near the ſea ſhore, beds of lava have been found at great depths, but how eaſily may have the mother ſtones of lava, hornblende, and ſhiſtoſe hornblende, be miſtaken for lava itſelf!

With reſpect to Ætna there can be no doubt. Dolomieu found immenſe heaps of ſea ſhells in its north-eaſt flanks at the height of near 2000 feet over the ſurface of the ſea. Hence he juſtly concludes that this volcano exiſted as a mountain before it was uncovered by the ſea; he adds, that at the height of about. 2400 feet there are regular ſtrata of grey clay filled with marine ſhells; theſe ſtrata muſt then have been depoſited while the mountain was a forming under the ſea; it contained alſo, he ſays, priſmatic lava, but the word *lava*, particularly with the addition *priſmatic*, can now impoſe on no one *.

He farther affirms, that in particular parts of this mountain, calcareous ſtrata exiſt under the lava.—So alſo Count Borch, in his Letters on Sicily and Malta, informs us, that the original ſtone of which Ætna conſiſts is granite mixed with jaſper, neither of which, ſurely, are lava; he ſays, that it abounds in *mines* of lead and copper, neither of which are ever found in lava, though their fragments may. This laſt mentioned geologiſt pretends that Ætna is at leaſt 8000 years old, which he infers from the beds of vegetable earth which he diſcovered betwixt different beds of lava. Yet Dolomieu expreſsly tells us that ſuch earth does not exiſt between beds of lava, *Ponces*, 472, and thus deſtroys the foundation of thoſe calculations that aſcribe to the

* Ponces, 465, 466, &c.

globe

globe an antiquity incompatible with the Mosaic history. Even if vegetable earth were found betwixt beds of lava, yet no conclusion relative to their age could fairly be deduced from that circumstance, as some lavas become fertile much sooner than others. Thus Chevalier Gioanni in 1787 found lavas projected in 1766 in a state of vegetation, while other lavas much more ancient still remained barren. Dolom. *Ponces*, 493. And in particular, it is well known that beds of volcanic ashes and pumice vegetate sooner than any other *.

I have been led into this detail by observing how fatal the suspicion of the high antiquity of the globe has been to the credit of the Mosaic history, and consequently to religion and morality; a suspicion grounded on no other foundation than that whose weakness I have here exposed. M. Dolomieu tells us, that Canon Recupero denied having ever expressed any doubt on that head, and could not conceive why a late celebrated traveller should endeavour to render suspicious the orthodoxy of his

* Ferber Italy, 169.

belief.

belief, So far from having been persecuted on that account, he had a pension from the court of Naples to his death, with many testimonies of esteem and regard. *Ponces*, 471.

ESSAY IV.

ON LAPIDIFICATION.

Any earthy ſubſtance whoſe integrant particles naturally cohere with ſufficient force to reſiſt the power of gravity, while one part of them only is ſupported and cannot be ſeparated by mere ſcraping with the nail, is called a *ſtone*; when they may be ſeparated by the nail, but not by an inferior force, they may be called *indurated earths*; but this being the limit, ſubſtances that thus cohere are alſo frequently denominated from either extreme, being ſometimes called earths, and ſometimes ſtones. When this coheſion is artificially produced, particularly by fire, they are called from the conſideration of other properties, *brick*, *porcelain*, *glaſs*, *&c.*

Hence the power of coheſion may be conſidered as the cauſe of induration; but this power itſelf is derived from two ſources, namely, the *general* attraction, or gravitation,

tion, of all particles of matter to each other *, and the *specific* attraction of the integrant particles of one species of matter, either to each other, or to those of another species; both these sorts of attraction are so much the stronger, and, consequently, so is also the resulting hardness, as the points approaching to contact are more numerous and nearer to each other † in the same mass, and these are capable of becoming so much the more numerous as the particles that present them are more minutely divided; their surface (relatively to their masses) being increased in proportion to their division.

Hardness is properly that sort of cohesion that resists division by abrasion, or scission, its opposite is *softness*. *Firmness* is that coherence which resists percussion, and its opposite is brittleness, or *fragility*. Brittleness arises from the elasticity of the particles struck, and may be possessed in a high degree by substances of great hardness, as is evident in glass, steel, bell-metal, &c. in

* Except those of the igneous element.

† For exact contact perhaps never takes place.

all

all of which the approximation to contact ſeems to be very near, but the points betwixt which this intimate approximation takes place are not very numerous, as appears by the low ſpecific gravity of glaſs; with reſpect to ſtony ſubſtances, the ſpecific attraction of ſiliceous particles, and alſo that of argillaceous particles to each other, ſeems to be the greateſt, that of calcareous ſubſtances next greateſt, and that of magneſian particles leaſt, in moſt inſtances.

Earthy ſubſtances acquire a ſtony hardneſs either from cryſtallization more or leſs perfect or confuſed, concretion, cementation, or ſubſtitution of unorganic to organic matter.

Each of theſe modes of Lapidification I ſhall now conſider.

Cryſtallization, when perfect, is an operation by which the component particles of bodies are ſo arranged in uniting to each other, as to aſſume a regular internal, and external form; to effect this arrangement they muſt be minutely divided, have liberty of motion, be placed at a due diſtance from each other, and be undiſturbed by a force ſuperior to that of their mutual attraction;

in proportion as theſe circumſtances more or leſs perfectly prevail or fail, the cryſtallization is more or leſs perfect or confuſed, as explained in the firſt volume of my Mineralogy, chap. 1. § 2.

As bodies may be minutely divided either by igneous ſolution, that is, fuſion, or by ſolution in a liquid menſtruum, cryſtallization may in appropriate circumſtances take place either in the dry, or in the liquid way; in the dry way, however, much more difficultly than in the liquid, becauſe the particles are too much crowded together; hence the moſt perfect cryſtallizations thus formed, are produced in the act of ſublimation, as I have often obſerved in expoſing different mineral ſubſtances to Parker's lens; and often alſo in crucibles, particularly with reſpect to magneſia, and ſtones of that genus. The cryſtallizations formed by mere fuſion are always imperfect, or rather rudiments of cryſtals, as the lamellated or ſtriated, or granular appearances of the different metals, and metallic ores.

But the natural cryſtals of ſtony ſubſtances were all (except a few found in lavas) formed in the moiſt or liquid way, no no known natural heat being ſufficient to produce their fuſion, and the circumſtances that accompany them being incompatible with igneous fuſion, as will be ſhewn in the laſt of theſe Eſſays. Nay, ſome cryſtals are found which by no poſſibility could be the reſult of previous fuſion, even though every other neceſſary circumſtance ſhould concur, namely, thoſe which with, or without an intermediate priſm, are terminated by a pointed pyramid at *both* ends, as thoſe of quartz and calcareous ſpar ſometimes are; for cryſtals formed by previous fuſion muſt neceſſarily adhere to ſome baſis in contact with them while in fuſion, elſe they could not be ſupported, but theſe could adhere to none without altering their ſhape.

That ſiliceous earth is ſoluble in ſimple water, when ſufficiently comminuted, appears from various obſervations. Mr. Genſanne in the mines of Cramaillot in Franche Compté, remarked that the water that tranſuded through the rocks that formed the vault of the works, produced concretions that reſembled ſtalactites, but were in reality quartz; they alſo ſometimes appeared

peared on the timber of the mine; this observation he attended to for ſeveral years. Hiſt. de Langued. vol. 2. p. 28, &c. 1 Buſſon Mineralogy, 48. Mr. De Laſſone found the ſurface of a ſandſtone which had the year before been inveſted with a ſiliceous cruſt, nearly as hard as agate, the particles of which it was formed muſt therefore have been conveyed and depoſited by water. Mem. Par. 1774, p. 13, in 8vo. Bergman found ſilex in the waters of Upſal, but in ſmall quantity; Klaproth in a much larger, in the waters of Carlſbad, 1 Klap. 335. 340. and though this water contains an alkali, yet it is not to this that the ſiliceous matter is indebted for its ſolubility, for the alkali is fully aerated; and Weſtrumb detected it in many more, as did Santi in the waters of Piſa. Nott's tranſlation, p. 53. The only queſtion is, whether it ſhould be ſaid to be mechanically ſuſpended? that it is truly diſſolved, appears to me moſt conſonant to truth; as Dr. Black, however, ſeems to think otherwiſe, any opinion oppoſite to that of a philoſopher of his acknowledged ability, deſerves, before it be admitted, the ſtricteſt examination;

the

the following proofs will, I hope, appear satisfactory.

1°. The Doctor himself found in an English gallon of Rykum water 21,83 gr. of siliceous earths, and 3 only of caustic natron; though caustic alkali has the power of dissolving siliceous earths, yet, surely, it cannot dissolve upwards of six times its weight of that earth, therefore, in this instance, the solution of so large a quantity of silex cannot fairly be attributed to the natron; among a variety of conjectures to explain this fact, the Doctor thinks it most probable that common salt and Glauber's salt had been applied to the earthy and stony strata, which contain mixtures of silex and argil; that these salts were in part decompounded by the attraction of these earths for the alkali of the neutral salt; that part of the acid had been dissipated or changed into sulphur and sulphureous gas by the simultaneous action of inflammable matter, and that the compound of alkali and earthy matter had afterwards been long exposed to the action of hot water. But this explanation the Doctor allows to be merely conjectural. I should think that if the ſiliceous matter had been decompoſed by its affinity to the alkali, the alkali would have ſtill adhered to it in the quantity neceſſary to hold it in ſolution, which we find it does not. Let this inſtance be compared with any other caſe of ſolution; when a metal or an earth is diſſolved by means of a menſtruum, if that menſtruum be withdrawn, or ſaturated with ſome other ſubſtance, does not the ſubſtance it had diſſolved immediately fall, unleſs the new compound be alſo a menſtruum for it? Thus, if magneſia be diſſolved in water by means of fixed air, or of a common acid, does it not fall as ſoon as the fixed air has evaporated, or as ſoon as the common acid is ſaturated with an alkali? But in the caſe before us the alkali is for the moſt part ſaturated, and yet the ſilex remains in ſolution for years: this inſtance is therefore of a totally different nature from that of the caſes adduced, and the ſolution muſt be attributed to the attractive power of the menſtruum that ſtill holds it in ſolution, namely, water.

In fact, the term, *ſolution*, denotes two different ſorts of action: firſt, that of the

menſtruum

menſtruum on the integrant or component parts of the aggregate to be diſſolved, whereby it ſeparates them from each other; and, 2do that by which it holds them in ſolution when ſeparated. The firſt ſort of action is that which the Doctor ſays he never obſerved water to poſſeſs with reſpect to ſiliceous earth, nor do I contend for it; the ſecond he certainly does not diſclaim expreſsly, and his own experiment proves it to exiſt in water; with reſpect to that earth, it is true, he ſeems to think this the effect of a mechanical ſuſpenſion, rather than that of a chemical attraction, becauſe he could never diſſolve flint, ever ſo finely pulverized, in mere water; but this argues only a defect of the firſt ſpecies of action, for ſurely the reaſon of this inſolubility is, that no artificial pulverization is ever ſufficiently minute. Thus we find, that argil once baked, is very difficultly, or ſcarce at all ſoluble in any acid, let it be ever ſo finely pulverized; and though acids be its natural menſtruums, yet if this argil be ſtill more ſubtilly divided, as it is by chemical agents in the act of precipitation, or even after precipitation, while ſtill moiſt, it is then eaſily ſoluble in appropriate acids; and to come cloſer to the caſe conteſted, the Doctor will allow that cauſtic alkali is the natural menſtruum of ſiliceous earths. Yet Mr. Macie obſerved, that powdered flints were ſcarce at all acted upon even by boiling fixed alkali, and the very little that was diſſolved was ſoon precipitated again in the form of minute flocculi on expoſure to the air (a proof that it was argil, and not ſilex); but the precipitate obtained from *liquor ſilicum* by marine acid diſſolved even when dry, and very readily in this alkali, and while ſtill moiſt did ſo very copiouſly, even without the aſſiſtance of heat *. It is, therefore, plain, that the inſolubility of ſilex in water, in common caſes, ſhould be attributed to the defect of a diviſion ſufficiently minute; the diviſion requiſite to render it ſoluble in water ſhould, perhaps, be ſtill more minute than that requiſite to diſſolve it in fixed alkalies, or rather the particles ſhould be ſtill more diſcrete, in order to enable them to be ſurrounded by a ſufficient quantity of water. Mechanical ſuſ-

* Phil. Tranſ. 1791, p. 385.

penſion

penſion is not very difficultly diſtinguiſhable from chemical ſolution. Filtration commonly ſeparates particles ſo ſuſpended, at leaſt if ſeveral times repeated, and the liquor gently heated; of this we have a remarkable inſtance in the caſe of iron precipitated by the Pruſſian alkali.

2do. Silex is found in the aſhes of all vegetables, as Mr. Bergman atteſts, § 172. *Anmerk* and Ruckert, but principally in the bambou reed, in whoſe joints even a pebble hard enough to cut glaſs has been diſcovered, *Macie's Memoir above quoted.* It is plain, then, that this earth is contained in the water imbibed by plants; the ſmall proportion of it that generally occurs in a given quantity of water, is no proof of a mere mechanical ſuſpenſion, for this may and ſhould be attributed to the rare occurrence of particles ſufficiently minute to be taken up by water; I ſay it ſhould be attributed to this circumſtance, becauſe, in ſome caſes, namely, where this circumſtance occurs, its proportion is very conſiderable: thus, Stucke found that according to his experiments 20 oz. of water collected in the internal cavities of baſaltic columns ſhould contain $14\frac{1}{7}$ grains of ſilex. In this caſe the water percolating the pores of the baſalt muſt have collected the minuteſt ſiliceous particles that occurred: here not a particle of alkali was found; but on the contrary, a large proportion of aerated magneſia and argil. *Stucke unterſuck*, 119.

3dly. Zeolite is alſo ſoluble in water, as Mr. Bergman has ſhewn. 3 Bergm. 255. Laſtly, to Dr. Black's authority I ſhall oppoſe that of a chymiſt equally reſpectable, that of Mr. Klaproth, who tells us, that from his own experience he has learned that in favourable circumſtances ſilex is ſoluble in water, without the co-operation of a fixed alkali, and that the hypotheſis of the Doctor to explain the ſolubility of ſilex, was no way requiſite. 2 Klapr. 108.

There are few examples, however, of the cryſtallization of ſtony ſubſtances at this day, the reaſon of which is very obvious: all theſe ſubſtances were originally created in that ſtate of minute diviſion which watry ſolution and cryſtallization requires; and the greater part of theſe have long ſince entered into a ſtate of combination or accretion, from which mere water can,

can, only in a great length of time, or perhaps never, diſengage very many of them. This remark is particularly applicable to the formation of cryſtals of the ſiliceous genus; of which genus, when pure and unmixed, 1000 parts water can take up only one, and whoſe perfect cryſtallization, moreover, requires perfect reſt, undiſturbed even by the alternate rarefaction and condenſation of the atmoſphere, as may be deduced from this circumſtance, that theſe cryſtals are always found in cavities well ſecured from the free communication of the air, as in the veins and cavities of mountains, or in hollow ſtones called *geods*, or in rifts, &c. where the air has had acceſs, or any diſturbance taken place, the cryſtallization is imperfect, being merely granular or diſtorted. Some have attributed the ancient ſolution of ſtones of the ſiliceous genus, to ſome imaginary menſtruum which, they ſay, has long ſince been deſtroyed or ſaturated. This ſuppoſition is both abſurd and gratuitous: abſurd, becauſe it is grounded upon another ſuppoſition, which evidently is ſo, namely, that ſiliceous ſubſtances were at firſt formed in

 a concrete

a concrete ſtate, that they might by this fictitious menſtruum be immediately after reduced to a diſſolved ſtate. Gratuitous, becauſe no trace of ſuch a menſtruum can be found; even the ſparry acid, the only known acid menſtruum of ſiliceous ſubſtances, has never been found in ſiliceous cryſtals, and the quantity of it known to exiſt, is infinitely too ſmall to effect ſuch a ſolution, and its affinity to ſiliceous earths is ſmaller than to earths of other genera, to which, conſequently, it would preferably unite. An alkaline menſtruum would be much more congenial, if any trace of it could be found.

In modern times it is only the ſiliceous particles that have eſcaped combination, or have, by ſome means, been detached from it, and often widely diſperſed, or thinly ſcattered through other ſtony maſſes, and ſlowly collected by the minute drops of water that circulate through theſe maſſes, that can in appropriate circumſtances form cryſtals, the drops of water gradually evaporating in theſe hollows, and depoſiting the ſiliceous ſubſtances, firſt on the baſis to which they adhere, and afterwards on each other.

other. The annual alterations of heat and cold which prevail even at the greateſt depths, though ever ſo ſmall, are ſufficient in a great length of time to condenſe, promote, and carry off, theſe vapours: this circulation of vapours in the interior parts of the earth, has been lately proved by Baron Trebra in his third letter.

That even ſiliceous cryſtals were formed in water, appears not only from the foregoing general reaſoning, but from various concomitant circumſtances. 1°. Baron Veltheim lately found ſome in a lonely retired ſpot, that ſeemed recently formed, being as yet ſoft. 1 Gerh. Geſch. 17. 2do. All of them, even the moſt ſolid and compact, loſe ſome part of their weight when expoſed to a ſtrong heat, and many of them decrepitate; the weight thus loſt is mere water. Thus zeolytes loſe from 5 to 18 per cent. as is well known; and in Klaproth's experiments compound ſpar loſes 45 per cent. opals from 6 to 18; ſhorl from 7 to 9; turmaline loſes 15 per cent. Braſilian topas 20; common flints 5; and red quartz 3 per cent. 1 Klapr. Beytr. 41 Roz. 95. Fleaurieu de Bellvieu found Carrara

rara

rara marble, heated below calcination, to loſe $\frac{1}{113}$ part of its weight.

3°. Pebbles filled with water ſometimes occur; nay, Ferber obſerved in the mineral cabinet of Piſa, a round quartz cryſtal, half filled with water, and even containing an inſect. If it be ſaid, the water was introduced through a chink, I aſk, how the inſect could be introduced? for the chink, if there were any, was ſo ſmall, that Ferber could ſcarce believe his eyes when he perceived the water, whereas the wonder would ceaſe if there were a perceptible chink *. Inſtances of the ſame kind have frequently occurred.

4°. Gerhard well remarks that ſtones formed in the dry way, being heated to redneſs, become ſtill harder, or at leaſt remain equally hard; but thoſe formed in the moiſt way being ſo treated, become ſofter: now ſiliceous cryſtals, thoſe of quartz for inſtance, become ſofter when heated to redneſs, therefore they muſt have been formed in the moiſt way † As to calca-

* *Italy*, 434, 21ſt letter.
† 2 Gerh. 118.

reous

reous ſpars, they have often been formed by art in the moiſt way, but, ſurely, never by igneous fuſion; Sauſſure obſerved ſome formed in a bottle of aerated water. 1 Sauſſ. 270. 2 Sauſſ. § 1097, in note.

5°. Siliceous cryſtals are found in the cavities of calcareous ſtones, where theſe are ſo ſituated that infiltrations from the former claſs may paſs into them; thus Mr. Sauſſure found cryſtals of quartz in a calcareous maſs of 400 feet in extent, which leaned againſt a mountain, formed of quartz and mica, on the ſide adjacent to this granilitic rock. 2 Sauſſ. 118.

6°. Siliceous ſtalactites have lately been found in Montarniata, hanging from Peperino, and alſo in the form of an incruſtation ſuperficially inveſting lavas; now this ſtone is infuſible in the ſtrongeſt heat, therefore it does not owe its origin to volcanic heat, but muſt have been formed by tranſudation, or infiltration through the lavas after they had cooled; in ſtructure, tuberoſity, rugoſity, &c. it exactly reſembles calcareous ſtalactites; ſee Chy. An. 1796, 589, &c.

Laſtly,

Laſtly, Siliceous ſolutions in fixed alkali, after the alkali has been ſuper-ſaturated with an acid, being ſlowly evaporated, depoſit the ſilex in the form of cryſtallized grains, which ſhews, that ſilex can cryſtallize in mere water, when the proportion of water neceſſary to hold it in ſolution is ſlowly diminiſhed. 1 Klapr. 211.

The next mode or immediate cauſe of Lapidification is concretion, that is, the cloſe union of earthy particles to each other, without any ſort of cryſtallization, but ariſing merely from their approximation to each other after the expulſion of the ſuperfluous water; thus clays are indurated, and many ſorts of ſtone of the argillaceous genus are formed, particularly when calces of iron, petrol, or carbonaceous ſubſtances are found in them; bricks have often been formed in this manner by mere ſolar heat in hot and dry climates. That the hardneſs thus reſulting is derived from their cloſer union and the expulſion of water, is clearly proved by the contraction of their dimenſions which they experience, and their loſs of weight; the Poliſh and Hun-

garian

garian huts formed of clay, are a further proof of the induration thus produced *.

We may, however, remark, 1°. that some proportion of water is always necessary to promote this lapidiscence, for earths that have all their water expelled, remain in dust, or if a considerable proportion be expelled, they remain much softer, and hence to harden them, some water must be added; this Dolomieu remarked with respect to lavas. *Ponces*, 417.—2°. That the dimensions of some compounds that acquire a stony hardness by concretion, are sometimes increased, namely, when they absorb air; the compound, for instance, that forms Pouzzolana expands while it hardens, as its ferruginous part absorbs the oxygen of water.

Water, in some small proportion, seems even an essential ingredient in many species of stone, even the hardest; quartz, for instance, loses its transparency when deprived of it. In these cases the water seems to be solidified by a loss of great part of its specific heat, in the same manner as that

* Schwed. Abhand. 1770, 195.

contained in Glauber's ſalt is now known to be.

The hardneſs induced by deſiccation in ſtones of the ſiliceous genus has been often remarked; thus, Delius obſerved, that Hungarian opals when firſt dug up, are ſo ſoft as to be friable betwixt the fingers; but by expoſure to the air and ſun only for a few days, they acquire a ſtony hardneſs. 44 Roz. 48. and Sauſſure tells us, it is well known to all mineralogiſts, that moſt ſtones, even granites, are harder on the ſurface than in the interior of mountains. 6 Sauſſ. 319. That many argillaceous ſtones are ſoftened by water, and hardened by expoſure to the air, is a matter of general obſervation. 2 Bergm. Journ. 1789. 724.

Even calcareous ſtones are hardened by deſiccation; this fact I often obſerved with reſpect to compact limeſtone taken from a quarry level with the ſurface of the earth, not only the ſuperior ſtrata, but even the inferior were much leſs hard while in the quarry than in a few days after they were taken out of it and expoſed to the air; this has often been obſerved by others. Mem. Par. 1746. 1075. 1 La Mether. 12.

That calcareous concretions of a ſtony hardneſs have been formed in modern times, and ſtill continue to be formed, particularly in the vicinity of ſprings ſtrongly impregnated with calcareous or calcareo ſulphureous matters, as in Derbyſhire, Bohemia, &c, is quite notorious; and that vaſt maſſes of limeſtone have within a few centuries concreted in the ſame manner, may be collected from the diſcovery of various artificial ſubſtances within thoſe maſſes. As theſe, however, have been by ſome, aſcribed to ſome fictitious Preadamtick periods, I ſhall quote one, which, without very violent ſuppoſitions indeed, cannot be attributed to any other but very modern times. In working a block of ſtone raiſed near Paris, the barrel of a piſtol was found imbedded in the midſt of it. 1 Buff. Mineral. 39. Stalactitic concretions of modern formation, and even artificial, are too well known to require any illuſtration; but are alſo a full proof of the formation of ſtones by concretion, or at leaſt, a commenced cryſtallization; and perfect

perfect calcareous ſpar has been found in a ſtalactitic form. 1 Bergm. Journal, 1792, 218.

The third cauſe of lapidification is *cementation*, ſee Black on Geyſer, p. 22, that is the introduction of particles, either of the ſame or of a different ſpecies, into the interſtices of ſubſtances that either did not adhere at all to each other previous to this introduction; or at leaſt were of a looſer or leſs indurated texture. Thus Sauſſure tells us, that in the neighbourhood of Meſſina, where grits are quarried near the ſea ſhore, the cavities formed by their extraction are ſoon filled with ſea ſand, which in a few years is ſolidified, having its particles agglutinated by the calcareous matter introduced by the ſea water. 1 Sauſſ.§ 305. p. 248. Bowles remarks, that in the neighbourhood of Cadix, the ſea poſſeſſes the ſame power, as the fragments of brick, mortar, &c. thrown on the ſhore, are, after a certain time, cemented with the ſand and ſhells into an uniform maſs of ſtone. Bowles' Spain, 99. Flurl relates, that fragments of rocks are cemented together, even at this day, by ſtreams impregnated with calcareous

careous matter, and depoſiting it on and between theſe fragments, at Hugelfing in Bavaria. Flurl Bavaria, 23, 24. Mr. De la Faille, an Academician of Rouen, obſerves, that the ſea near Chatelaillon after a ſtorm throws up a ſort of mud on which, after a few days, a ſpecies of ſhell-fiſh, called griffites, appear, and ſoon after, the whole hardens into a ſtone as ſolid as the hardeſt limeſtone. 20 Roz. 43, in note*. But, perhaps, no where does this effect take place ſo quickly as at Crainburg, near the banks of the Save; for in quarrying a ſtratum of ſtone, it was obſerved, that the fragments, if not immediately removed, were in a ſhort time ſo firmly cemented by the river water that oozed through the banks, that they required to be quarried over again. 2 Born Phyſ. Arbeit 8vo. in note; and hence the ingenious author, Mr. Gruber, clearly deduces the origin of breccias. In many caſes calces of iron minutely divided form the whole of the cement; or at leaſt powerfully contribute to the cementing power of other earths. Zimmermann

* See alſo Mem. Par. 1721. 343, 8vo.

mixed one part filings of iron and three parts ſand, ſprinkled, or rather covered them with water, and let them ſtand ſix months, at the end of which period he found the veſſel burſt by the expanſion of the oxygenated iron, and the ſand ſo firmly compacted, that the maſs thus formed could not be broken but by a chiſel and hammer. Henckel *Origine des Pierres*, 405 in note; and that this induration may, and does take place at great depths in the ſea, is evidently proved by the obſervations of Rinman. Mem. Stock. 1770, related by Gadd, that an iron anchor long depoſited in the ſea, had hardened into ſtone all the ſand, clay, and ſhells, which ſurrounded it, to a pretty conſiderable diſtance; and is farther confirmed by a ſimilar obſervation of Mr. Edward King, Phil. Tranſ. 1779, p. 35, that a violent ſtorm having laid bare part of the wreck of a man of war that had been ſtranded 33 years before, ſeveral maſſes, conſiſting of iron, ropes, and balls, were found covered over with a hard ſubſtance which upon examination appeared to be ſand concreted and hardened into a kind of ſtone: that which concreted round

the

the rope, retained the impreſſion of that part of the ring to which the rope was faſtened in the ſame manner as the impreſſions of extraneous foſſils are often found in various ſtrata. Alſo round the iron hand'e of a braſs cannon that remained in the ſea a much longer time, a much harder incruſtation of ſand was found, in cloſing cockles, muſcles, limpets, oyſters, &c. all ſo firmly fixed, and converted into a ſubſtance ſo hard, that it required as much force to ſeparate or break them, as to break a fragment of any hard rock. Ib. 40, 41. It appears alſo, that a very ſmall proportion of calx of iron is ſufficient to produce induration, when diffuſed through the maſs of earthy matter, not only by the obſervation of Rinman above related, but alſo by that of Mr. King on the induration cauſed by the point of a nail, in the paper above quoted.

Stones already formed, may be ſtill further indurated by the infiltration of ſlightly oxygenated iron; thus Dr. Fothergill having watered pieces of Portland ſtone with water impregnated with iron ruſt, found it in a few years to have acquired a ſenſible degree of ſuch hardneſs as to yield a metallic ſound, and reſiſt any ordinary tool. Phil. Tranſ. 1779, 44.

That a proceſs of the ſame nature has been, and ſtill is, carried on in the interior parts of the earth, wherein hardneſs is induced or increaſed by the infiltration of particles of the ſame, or of a different, nature, appears by many obſervations, of which I ſhall only mention a few, as being perfectly deciſive.

1°. Mr. Werner obſerved in the mountain of Zeigelberg, ſtrata of blue clay and compact red iron ſtone, to alternate ſeveral times with each other, and that each had peculiar petrifactions not found in the other; for inſtance, only turbinites were found in the one, and in the other only chamites, or muſcullites, &c. which ſhews theſe depoſitions were not ſimultaneous, but originated at different periods of time. Now the petrifactions found in the ſtratum of the blue clay, when placed under the iron ſtone, were conſtantly filled with the ferruginous matter of that ſtone, and, on the contrary, the petrifactions found in the iron ſtone, when under the ſtratum of clay, were

were filled, at leaſt in the parts contiguous to the clay, with argillaceous matter, but the hardneſs of the iron ſtone prevented it from receiving much of the argillaceous matter. Wedem. Umwandl. p. 118.

2°. The petrifactions found in chalk are frequently filled with ſiliceous matter, and ſo hardened as to give fire with ſteel. Now chalk itſelf is found conſtantly to contain more or leſs of minute ſiliceous particles; it is, therefore, to the infiltration of theſe that this increaſed hardneſs is to be attributed.

3°. It is remarked, that coloured marbles (of the compact kind) are generally harder than the white, as by infiltration or otherwiſe they received, either during or after their formation, particles of another ſpecies, by which their interſtices were filled. Nadault. 1 Buffon Mineral. 342.

4°. The calcareous Farcilite called Amenla, is formed of rounded calcareous maſſes of extreme hardneſs, cemented by a calcareous cement. Now near the ſurface where the calcareous cementing matter could not be ſo abundant, theſe ſtones are but looſely connected or united, but at greater depths, for the contrary reaſon (there being more calcareous matter incumbent) the cement becomes ſo hard that the maſſes it forms cannot be ſeparated but by exploſion with gunpowder; per Sauvages Mem. Par. 1746, 1086, in 8vo.

That the cementing matter was of poſterior formation, is evidently inferred from the veins of calcareous ſpar that run through the rouned maſſes, but which never paſs into the cement, as the ſagacious author well remarks, p. 1091.

5°. Guſman atteſts, that he has ſeen petroſiliceous breccias, whoſe calcareous cement originated from their own decompoſition. Lithophy. Mitſian. 114.

6°. The refuſe ſtones thrown out of mines are frequently hardened again by the matter ariſing from their diſintegration and decompoſition, as Laſius obſerved at Ramelſberg. Hartz. 283 and 3. De Luc Lettres à la Reine, 298. Sometimes a commenced cryſtallization is found in theſe adventitious accumulations. Flurl *Bavaria*, 565.

It ſometimes happens that a cement which was originally a mere confuſed aggregation, becomes, after ſome ages, ſo minutely comminuted by ſolution, as to cryſtallize into a tranſparent matter; this, Mr. Nadault obſerved in the fragments of a very ancient rampart, in ſeveral fathoms of which the ſtones were connected, not by mortar, but by a tranſparent ſubſtance (a ſpar) into which the calcareous particles of the mortar were converted by the infiltration of rain water *; the ſame effect ſeems alſo naturally to take place in many limeſtones of a looſe ſpungy texture, the water that pervades part of their maſs gradually conveying into another part of the ſame maſs ſuch particles as it can diſſolve: theſe being thus brought into a cloſer union with each other, cryſtallize, and induce an additional hardneſs in that part of the maſs in which they ſettle. The hardneſs thus produced is the reſult, not of one, but of ſeveral ſucceſſive infiltrations, and hence various degrees of hardneſs are obſerved to take place between the looſe and ſpungy extremity of the ſtone, and the extremity which has received its maximum of ſolidity.

* Per Nadault, 1 Buffon Mineral. 391.

Calcareous maſſes of this ſort have been obſerved by the Abbé Sauvages* and others. In Mr. Greville's cabinet there are many hornſtones, in which the gradual tranſition from an indurated clay into ſiliceous hardneſs and fracture, may be evidently diſcerned. It has been aſked, what becomes of the water that conveys theſe cryſtallizing particles? The anſwer is eaſy; all ſtones, and even the denſeſt metals, contain vacuities; theſe act as capillary tubes, and ſoon reconvey the water to the upper ſurface, from which it gradually evaporates.

Of the hardneſs induced by obſcure and confuſed cryſtallizations pervading ſtony maſſes, we have an indubitable proof in many impure gypſums, which undoubtedly coaleſced by cryſtallization, though the cryſtalline grains are ſcarcely diſcernible.

The cements are of many different ſorts, but more generally of the ſame ſpecies as the ſtone cemented. Sometimes they run in veins; ſometimes they are diſperſed through the whole maſs inviſibly to the

* Mem. Par. 1746, 1105, in 8vo.

naked

naked eye, or viſible only in a ſtrong light, or by a lens; ſometimes very viſible in minute ſhining ſpecks: thus, in the calcareous claſs they are more frequently *ſparry*, but often alſo ſiliceous or ferruginous; ſometimes argillaceous or pyritous, more rarely gypſeous or fluoric—in the argillaceous they are commonly ferruginous, but ſometimes calcareous or ſiliceous, &c. and, as lately has been diſcovered, ſometimes carbonaceous.

The laſt known mode of inducing ſtony hardneſs, is SUBSTITUTION; that is, the introduction of ſtony, and, ſometimes, of metallic ſubſtances, into organic bodies whether of the vegetable, or of the animal kingdoms, in proportion as the particles of theſe organic ſubſtances are deſtroyed by putrefaction, ſo as to aſſume the place, and, conſequently, the form and figure of theſe, as if caſt in the ſame mould.

The mineral ſubſtances, thus moulded, are called *petrifactions* in the moſt proper ſenſe of the word; but, by many, particularly by the Germans, this word is uſed in a looſer ſenſe, to denote any organic ſubſtance found buried at great depths in the earth,

earth, or embodied in ſtone, whether converted into ſtony matter or not.

The hardneſs thus induced, ariſes from the proximity of the ſtony particles to each other, and the ſuperior attraction of theſe particles, in compariſon of thoſe of the organiſed ſubſtances whoſe place they occupy.

The mechaniſm of petrifaction, I conceive to be ſhortly this, organic ſubſtances petrified, are either found in water, replete with the ſtony matter found in petrifactions, or, they are incloſed in earths, ſands, or ſtones; in both caſes the firſt ſtep in the proceſs of petrifaction, is the eſcape of the hydrogen and part of the carbon of which the organiſed ſubſtance conſiſts. If this ſubſtance be ſurrounded by water ſtrongly impregnated, the ſtony particles are immediately attracted and ſubſtituted in the place occupied by the particles that eſcaped; but if the petrifaction takes place more quickly than water can ſupply the ſtony particles, then no petrifaction takes place, and hence the ſofter organiſed ſubſtances are ſcarce ever found petrified. The proceſs is thus continued by gradual

putrefaction

putrefaction on the one part, and gradual insinuation of stony particles on the other, until the petrifaction is completed.

Shells, bones, and woods, are the substances that decay most slowly, and therefore are most frequently found petrified; but the fibrous parts of these decay more slowly than the softer or medullary, and thence the disposition of the stony particles introduced, must, necessarily, be as different as the disposition and form of the fibrous particles were, whose form and place they assume. We must also conceive that the petrifaction is at first imperfect, and not absolutely completed until long after it has penetrated into the interior of the organic substances. The minute interstices at first left, are afterwards gradually filled up, though in some instances contrary appearances occur.

In some cases, also, the interior or more central parts of the organised substance first decay, while the exterior remains sound; in such cases the petrifying operation takes place only in the interior: this has often been observed in woods where the wood is of a species that strongly resists putrefaction, or or the water in which it is lodged is but ſlightly impregnated with petreſcent particles, the petrifaction very ſlowly takes place; of this we have a memorable inſtance in one of the timbers that ſupports Trajan's bridge over the Danube, ſome miles below Belgrade. About the year 1760, the emperor of Germany being deſirous to know the length of time neceſſary to complete a petrifaction, obtained leave from the ſultan to take up and examine one of theſe timbers. It was found to have been converted into an agate, to the depth only of half an inch; the inner parts were ſlightly petrified, and the central ſtill wood. Undoubtedly the timber employed was of the kind leaſt ſubject to rot, and the Danube is not known to contain any notable quantity of ſiliceous particles; but the fact is important, as it proves to a demonſtration, that ſiliceous particles are ſoluble in water, are taken up by wood, that petrifactions are carried on in appropriate circumſtances in modern times, and the ſucceſſive proceſs of petreſcence as above ſtated. Juſti Geſch. des Erdkorpers, 267. 1 Gerh. Geſch. 222.

But in the moſt favourable circumſtances; that

that is, where the wood is of a ſpecies more ſuſceptible of putrefaction, and the water in which it is immerſed, richly impregnated with ſtony particles, petrifaction takes place much more quickly. Don Ulloa tells us, that north of Quito, at the foot of mount Anlagua, there is a river that petrifies any ſort of wood or leaves that are thrown into it, and that he had whole branches thus petrified; the fibres of the rind, even the ſmalleſt fibres of the leaves, and the meanders of the fibrillæ being equally diſcernible as when freſh cut from the tree.

All the rocks in this river, he obſerves, are covered with a cruſt little inferior in hardneſs to the rocks themſelves. This matter, he adds, faſtens much more eaſily on corruptible ſubſtances, and frequently forms a lapideous tegument round the leaves, &c. * Mr. Stedel found the pieces of elm left in a fountain, near Ulm, become petrified in ſeven years †.

The petrifaction induced in woods ſurrounded by ſands, or incloſed in ſtones,

* Ulloa's Voyages, p. 377, Iriſh edition.
† 6 Roz. 8vo. 3d part, p. 18.

originates

originates exactly from the same causes, and is produced in the same manner. Mr. Gledisch observed one of the roots of a pine tree still in a state of vegetation, converted into the calcareous petrifaction called Osteocolla: it was surrounded by sand, and part of the wood in a rotten state remained in the center of the root. Mem. Berl. 1748, p. 49, 50.

This same petrifaction, which exactly resembled the root of a tree, Margraff analysed, and found it to yield volalkali from some remains of the putrid wood that were still contained in it. 1 Margr. 246. 261; and that putrefaction ever precedes petrifaction, may be deduced from the existence of fixed vegetable alkali in the marls that surround petrifactions, and the volalkali also often obtained by distilling them; see Gesner's Dissertation in 6 Roz. in 8vo. 2d part, p. 20, &c.

Several lakes or other waters that anciently possessed a petrifying power, have since lost it by having imparted the greater part of the stony particles they contained to such substances as were capable of retaining them.

ESSAY

ESSAY V.

ON THE DECOMPOSITION AND DISINTEGRATION OF STONY SUBSTANCES.

Decompoſition conſiſts in the ſeparation of the *conſtituent* parts of a ſtone or other ſubſtance, and may be either total or partial. *Diſintegration* denotes the ſeparation only of the *integrant* parts; both often take place in the ſame ſubſtance.

The only cauſes of *mere* diſintegration as yet known, are the viciſſitudes of the atmoſphere; the abſorption and congelation of water; the ſudden dilatation or contraction produced by the former, particularly when extreme, cannot but looſen the texture of moſt ſtony ſubſtances, and when aided by the abſorption of water, ſtrongly tend to ſeparate them. The water thus received in their minuteſt rifts, being afterwards frozen, burſts them with incredible force, of which frequent inſtances occur in the northern countries, and in the more elevated

vated mountains of the ſouthern, where the moſt ſudden tranſitions of heat and cold, and the higheſt degrees of the latter frequently prevail; and hence the broken craggy ſtate of their loftieſt ſummits.

The known external cauſes of decompoſition, are water, oxygen, and fixed air.

The internal cauſes are, the baſes moſt capable of forming an union with the external, as ſaline ſubſtances, ſulphur, ſlightly oxygenated calces of iron, or of mangaueſe, lime, argil, bitumen, carbon, and mephitic air, which is certainly contained in many ſtony ſubſtances, as Dr. Prieſtley has ſhewn in the firſt volume of his laſt edition, p. 64; but as to its nature and effects it is at preſent too little known: all theſe are aſſiſted by a looſe texture, of the ſubſtance acted upon.

Saline ſubſtances, particularly when (relatively to their maſs) they preſent a large ſurface, are diſſolved by water, and, conſequently, the ſtones, of which they ſometimes form a compenent part, are decompoſed; thus muriacite, which conſiſts of 27 per cent gypſum, 14 common ſalt, 5 mild calx, and 53 of micaceous ſand, muſt be be decompoſed when long ſubjected to the action of water.

Sulphur promotes decompoſition by abſorbing oxygen, while it is thus converted into vitriolic acid; but moiſture is alſo requiſite. To this cauſe the decompoſition of ſuch ſtones as contain pyrites is to be attributed; it ſeldom acts, however, unleſs united to ſome metallic ſubſtance; and hence its combinations with argil, unleſs aſſiſted by heat, are not ſenſibly decompoſed, or only in a great length of time.

Calces of iron, moderately oxygenated, are the moſt general cauſe of decompoſition, particularly when aſſiſted by a looſe texture, and the other cauſes of diſintegration; theſe act by abſorbing a greater proportion of oxygen and fixed air, but require alſo the aſſiſtance of moiſture. By this abſorption they gradually ſwell, and are diſunited from the other conſtituent parts of the ſtone into whoſe compoſition they enter.—When leaſt oxygenated their colour is *black*, or *brown*, or *bluiſh*; and in ſome inſtances, when united with argil and magneſia, *grey*, or *greeniſh grey*; the former in proportion as they become more oxygenated

 become

become purple, red, orange, and, finally, pale yellow; the latter become at first blue, then purple, red, &c.

Iron, in its perfect metallic state, or at least but slightly oxygenated, also decomposes water; but if exposed to the air it becomes farther oxygenated, and the compound in which it enters gradually withers, as Dr. Higgins observed, in imitating pouzzolana, on Cements, 124.

But stones into whose composition calces of iron highly oxygenated seem to have originally entered, are very difficultly decomposed, as red jaspers, &c. as they already possess nearly as much as they can absorb.

Manganese, when slightly oxygenated, is known to attract oxygen strongly, particularly with the assistance of heat and moisture; hence it is, in many cases, a principle of decomposition, as in sidero calcites, &c. it also frequently assists or promotes that effected by calces of iron.

Lime, from its attraction to fixed air, and its solubility in water, must promote, in favourable circumstances, the decomposition of stones, of which it forms a constituent

ſtituent part; to it the decompoſition of felſpars, and many zeolites, may, in part, be attributed *.

Argil, when its induration does not exceed 7, muſt, by the common annual viciſſitudes of heat and cold, gradually become rifty, abſorb, ſoften, and ſwell, and thus promote diſintegration, and decompoſition.

Bitumen is ſaid to form the cement of ſome limeſtones, and, probably, of various other ſpecies. Bowles found it ſo in various parts of Spain, and Flurl in Bavaria; and to its fuſion and withering (probably by attracting oxygen) he attributes the diſintegration of ſeveral compact limeſtones in Bavaria. p. 78.

Carbon has lately been found in ſeveral ſpecies of ſtone; as it powerfully attracts oxygen, to it we may, perhaps, attribute the diſintegration of many of them, as marls, marlites, ſome, argillites, ſhales, &c.

Mephitic air (the azote of the French) by its property of forming nitrous acid, when, during its naſcent ſtate, it is gradually

* That the calx is in a cauſtic ſtate, ſee Pelletier's analyſis. 20 Roz. 422. § 7 and 17.

brought into contact with the oxygen of the atmosphere, in a moderately dry state, may also promote decomposition; calcareous stones are known to contain it in pretty considerable proportion, and those that contain animal remains, probably, most; from this consideration we may derive some explanation of a very remarkable phenomenon related by Mr. Dolomieu. 36 Roz. 116. " All the houses of Malta are built of a fine " grained limestone, of a loose and soft " texture, but which hardens by exposure " to the air. There is a circumstance " which hastens its destruction, and reduces " it to powder, namely, when it is wetted " by sea water; after this it never dries, but " is covered by a saline efflorescence, and a " crust is formed some tenths of an inch " thick, mixed with common salt, nitre, " and nitrated lime; under this crust the " stone moulders into dust, the crust falls " off, and other crusts are successively form- " ed, until the whole stone is destroyed. " A single drop of sea water is sufficient to " produce the germ of destruction; it forms " a spot which gradually increases and " spreads like a caries through the whole

" mass

" maſs of the ſtone ; nor does it ſtop there, " but, after ſome time, affects all the " neighbouring ſtones in the wall. The " ſtones moſt ſubject to this malady are " thoſe that contain moſt magneſia ; thoſe " which are fine grained, and of a cloſe " texture, reſiſt moſt." Short as this account is, it appears from it, that the limeſtone of Malta contains both calcareous earth and magneſia, but moſt probably in a mild ſtate ; and the ſtone being of the looſer kind, is of the ſpecies which is known to contain moſt mephitic air. Mr. Dolomieu ſhews at the end of his tract on the Lipari iſlands, that the atmoſphere of Malta, in ſome ſeaſons, when a ſouth wind blows, is remarkably fouled with mephitic air, and at other times, when a north wind blows, remarkably pure ; and hence, of all others, moſt fit for the generation of nitrous acid.—Again, ſea water, beſides common ſalt, contains a notable proportion of muriated magneſia, and a ſmall proportion of ſelenite. From theſe data we may infer, that, when this ſtone is wetted by ſea water, the ſelenite is decompoſed by the mild magneſia contained in the ſtone, and

 intimately

intimately mixed with the calcareous earth; of this decompoſition, two reſults deſerve attention, 1. The production of vitriolic Epſom; 2°. The extrication of mephitic air, the muriated magneſia of the ſea water ſerving, during this extrication, the purpoſe of attracting and detaining a ſufficiency of moiſture. This air, thus ſlowly generated, and meeting the dry oxygen of the atmoſphere, forms nitrous acid, highly mephitiſed, but it ſoon acquires a due proportion of oxygen by deoxygenating the vitriolic contained in the Epſom ſalt, which by ſucceſſive depredations of this ſort is gradually deſtroyed. Part alſo muſt unite to the mild calx, which in its turn is decompoſed by the remaining mild magneſia; more mephitic air is ſet looſe, and more nitrous acid is produced, until the ſtone is deſtroyed; how the alkaline part of the nitre, which is one of the products reſulting from the decompoſition of this ſtone, is formed, is as yet myſterious; Is it not from the tartarin lately diſcovered in clays and many ſtones? I am as yet inclined to think that it is derived from the putrefaction of vegetable and animal ſubſtances; and

and though nitrous acid formed of oxygen and air, from putrefying ſubſtances, be found united, not only to the abſorbent earths to which it is expoſed, but alſo to a fixed alkali; yet I ſhould rather ſuppoſe that the alkali is conveyed into thoſe earths by the putrid air, than newly formed; and the reaſon is, that tartarin, notwithſtanding its fixity, is alſo found in ſoot, and in the ſame manner may be elevated in putrid exhalations. As to the common ſalt, ſaid alſo by Dolomieu to be found in the bliſters of this mouldering ſtone, I am as yet in doubt, for common ſalt was alſo ſaid to accompany the native nitre found in the pulo of Appulia, yet Klaproth in analyſing this nitrated earth could find none; ſee Zimmerman's account of this native nitre. 36 Roz. 111. 113, and 1 Klap. 319.

So alſo when the calx of iron contained in ſtones is but ſlightly oxygenated, it may, by reaſon of the cloſe texture of the ſtone, remain undecompoſed for ages; but if by any accident, as fracture, or contact with ſome ſaline matter, or the alternate reception and diſmiſſal of water, the reception of more oxygen is facilitated, a decompo-

 ſition

ſition will commence, which, as in the former caſe, will ſpread like a caries, becauſe the leſs oxygenated part of the iron takes oxygen more eaſily from the more oxygenated part, than from the atmoſphere, by reaſon, that the abſorbed oxygen is more condenſed than it is in the atmoſphere. Thus, iron inſerted into a highly oxygenated ſolution of vitriol of iron, and which, therefore, refuſes to cryſtallize, will take up the exceſs of oxygen, and thus reſtore the ſolution to a cryſtallizable ſtate; or as calx of tin takes up oxygen from calces of ſilver, antimony, &c. in the beautiful experiments of Pelletier. 12 An. Chym. 229, &c.

Hence, alſo, ferruginous ſtones near, or upon, the ſurface of the earth, being more expoſed to air and moiſture, and the diſruptive action of growing vegetables whoſe roots pierce through their minuteſt rifts, and, by ſwelling, burſt them, are more expoſed and ſubject to decompoſition. Water carries down the ferruginous particles into the lower ſtrata, and forms there thoſe illinitions and maſſes of piſiform argillaceous iron ore, which Buffon and others have, without

without ſufficient reaſon, derived from decayed vegetables *.

Baſalt, when pure, ſtrongly reſiſts decompoſition, or its ſurface alone bears any marks of it; the argillaceous, ſiliceous, and calcareous ingredients, and part of the ferruginous, ſoon recombining and forming a hard cruſt, which inveſts and protects the remainder of the ſtone;—but wacken is very eaſily decompoſed, and hence the baſalts or traps into whoſe compoſition it enters, yield eaſily to the decompoſing principle. Some granites, I may ſay moſt, are in appropriate circumſtances not difficultly decompoſed; the mica and felſpar are chiefly affected; the ſame may be alſo ſaid of moſt ſandſtones, particularly thoſe whoſe cement is argilaceous or ferruginous, and many porphyries, and gneiſs.

* See alſo Flurl *Bavaria*, 191, 192.

OF MOUNTAINS.

Among the various inanimate objects which Nature has ſo profuſely ſcattered around us, there are none which at firſt ſight convey ſo awful an impreſſion of the power of its great Author, as thoſe ſtupendous maſſes we call Mountains; none in which reflection diſcovers more convincing proofs of wiſdom and beneficence, than in their diverſified heights and arrangement, exactly ſuited to the varieties of their geographic poſition and the general economy of the globe. Without them the earth would be little more than a ſandy deſert, and the atmoſphere a peſtilential receptacle of noiſome exhalations; by conducting the electrical fluid, and the principle of heat, they contribute to the production of rain, which fertilizes the former, and purifies the latter. Their elevation enables us to extract metallic, combuſtible, ſaline, and other ſubſtances, whoſe uſe is indiſpenſable, yet which in *flat* ſituations, from the impoſſibility of drawing off the water, we could not obtain. Among the ſtony ſubſtances they

7 preſent

present us, many are applied to building, and to various arts; many are the harbingers of metallic or other valuable substances, and many others, both stony and metallic, exist, whose uses, through the unpardonable neglect of former ages, are as yet unknown; mankind unacountably forgetting that the principal occupation originally assigned to them was to *cultivate*, that is, to labour on, and extract every possible advantage from the earth, and the substances it contains. Nor is the wisdom of the geographic position of mountains, and of the degrees of their diversified elevation, suited thereto, less obvious and striking; thus, in the north-east parts of our continent, the vast Asiatic platform, from which so many mountainous chains branch forth, afforded, in the infancy of the globe, an habitation for man and animals; while inferior regions, for the purpose of completing their arrangement, still remained buried in the bosom of the deep. The height of these mountains that raise their lofty summits in the eastern parts, is proportioned to the course which their mighty rivers must hold in the extensive empires of Indostan and China, and fitted

fitted to produce the refreſhing blaſts neceſ-ſary to moderate the ardour of thoſe ſultry climates; whereas, in the more weſtern tracts, the ſame reaſons not exiſting, the elevations are far leſs conſiderable. In the ſouthern parts of Europe, the accumulated and exalted maſſes of the Alps, Appenines, and Pyrenees, diſpenſe the ſame bleſſings as in the north-eaſt part of Aſia; and on the other hand, in Africa and Arabia, immenſe ſandy plains occur, whoſe heated ſurface produces thoſe alternations of atmoſpheric currents that occaſion the monſoons, and the varieties of ſeaſon requiſite for the fertility of the tropical regions.

In common language, mountains are diſtinguiſhed from hills only by annexing to them the idea of a ſuperior height, not aſſigning to either the exact height that ſhould entitle it to its particular denomination. Geologiſts have aimed at greater preciſion; Pini and Mitterpachter call any earthy elevation a *mountain* whoſe declivity makes with the horizon an angle of at leaſt 13°. and whoſe perpendicular height is not leſs than $\frac{1}{5}$ of the declivity. Mitterp. 182.

Werner calls a mountain *high*, when its perpendicular

perpendicular height exceeds 6000 feet; *middlesized*, when its height reaches from 3000 to 6000 feet; and *low*, when its height is beneath 3000 feet. Berg. Kal. 176. Betwixt the tropics, the boundaries of vegetation are fixed at the height of about 12000 feet; in the temperate climates at from 5 to 8000; and within the polar circle still lower.

Before I proceed farther, I must notice an ill-founded opinion, advanced by a late highly respectable philosopher, the reverend Doctor Michel; Transf. for 1760, vol. 51, p. 584. namely, that they were forced up from the earth, and, consequently, not formed by precipitation from a fluid, as I have stated in my first Essay; this notion was suggested to him "by remarking," he says, "that, in all high and mountainous "countries, the strata lie in a situation more "inclined to the horizon than the country "itself; the mountainous countries being "generally, if not always, formed out of "the lower strata of the earth—from this "formation of the earth, it will follow, "that we ought to meet with the same "kinds of earths, stones, and minerals, ap-

" pearing at the ſurface, in long narrow " ſlips, and lying parallel to the greateſt " riſe of any long ridges of mountains; and ſo " in fact we find them." The only proofs he gives of this diſpoſition are, " that the " Andes have a chain of volcanoes extend- " ing 5000 miles, which are all, *probably*, " derived from the ſame ſtratum; and that " another chain, parallel to theſe, runs at " leaſt 100 leagues; and that the gold and " ſilver mines worked by the Spaniards, " are found in a direction parallel to theſe." It muſt be evident, that here is not even the ſhadow of a proof, not a ſingle ſtratum common to the plains and mountains is mentioned, much leſs a ſucceſſion of ſtrata, which alone could afford a proof. The Andes and American mountains were moſt injudiciouſly choſen, as, in reſpect to their compoſition, they are perhaps of all others the leaſt known, having never been deſcribed by any mineralogiſt; nor has the ſtratum, on which inflammation depends, in any volcano been ever known by obſervation. But of the very reverſe of his opinion, numberleſs inſtances might be adduced. Though the ſtrata of mountains

are

are often inclined to the horizon; yet many are perfectly horizontal, as will be seen in a subsequent Essay. I shall at present quote only one instance, namely, that of the mountain of Kinneculla in West Gothland; it consists of five different strata, *all* horizontal, the last reposes on granite; and of this, and this only, the neighbouring plains consist: the four others are found only in the mountain. Now by Dr. Michel's account, all of them should be found in the plains, and the stratum that is highest in the plain, should also be highest in the mountain; whereas the only one common to the plain and mountain, is that which is lowest in the mountain. 29 Swed. Abhandl. 24. 5 Bergm. 115, 128. 3 Bergm. 214. It rarely happens, that the strata of mountains conform to their convexity, and bend into the valleys. Mem. Par. 1747, 1082 in 8vo.

Mountains are said to have their *course* in that direction of their length in which they descend, and grow lower, or in the direction of the stream of a river, when any runs parallel to them. The course of mountains is seldom uniform. Bourguet and Buffon pretend that in two parallel chains,

chains, the ſaliant angle of the one conſtantly correſponds with the internal angle of the other, but ſubſequent Geologiſts utterly deny this correſpondence, except where a river runs between them. Sauſſure and De Luc deny it to take place in the Alps. 1 Sauſſ. 402, 411. 2 Sauſſ. § 920. Fortis in Dalmatia, 459. Pallas in Siberia, Act. Petrop. p. 40, and 1 Nev. Nord. Betyr. 294. Gentil in Eaſt Indies, Mem. Par. 1781, p. 433; hence the fantaſtic though beautifully decorated theory of Buffon, reſting principally on this foundation, falls to the ground.

As not only groups of mountains, but even ſingle mountains are formed of various materials, their claſſification cannot be deduced from the nature of ſubſtances thus variable and diverſified. It muſt then be founded on ſome general relation of the maſſes of which they conſiſt, with other ſubſtances foreign to the mineral ſyſtem; and of theſe relations, the moſt general, and to which all other properties of mountains are eaſily referable, is that of their priority, or their poſteriority, to the exiſtence of organized ſubſtances. Hence their

primary

primary divifion is into *primeval* and *fe-condary* or *Epizootic*. And the epizootic mountains are ftill farther diftinguifhable into *original* and *derivative*. The clafs of fecondary, and, perhaps, alfo that of primary, may be fubdivided into inert and ignivomous, into volcanic, and pfeudo volcanic. The volcanic have indeed hitherto been generally referred to the fecondary mountains, but as feveral of the Andes are faid to be volcanic whofe height exceeds that of any known fecondary mountain, for inftance, Catopaxi, and as the materials of volcanic mountains have been found in fome primary mountains, it is highly probable, that primary mountains alfo may fometimes be the feat of volcanos, hence I fhall treat of thefe apart.

The *moft extenfive* montanic ranges commonly confift of three chains, of which the internal are generally *primary* and the external *fecondary*; the internal is generally narrow and often fharp, the external broader and more extenfive. See Pallas Act. Petropol. Vol. 1. 30. Gentil. Mem. Par. 1781, 433. 1 Sauff. 189. 30 Roz. 275. 39 Roz. 401. Defcrip. Pyren. 144.

Some mountains diverge from a high extenſive platform, as the numerous chains that ſhoot from the Altaiſhan Platform. Pallas ibid. Others ſhoot like branches, from ſome conſiderable trunk, others appear retiſorm, croſſing each other in various directions, and ſome few ſtand ſingle.

Many mountains are ſteep on one ſide and gently inclined to the plains on the other: the ſteepneſs often ariſes from the rupture of the ſtrata, often from their decompoſition, being more expoſed to rain and impetuous predominant winds on one ſide than on the other. The gentle inclination often proceeds from the unequal extenſion of the ſtrata, the lower being the moſt extenſive, and the higher gradually narrower; often alſo from the failure and depreſſion of the lower ſtrata. The cauſes of which are to be ſought for in the natural hiſtory of the different countries in which they occur.

The appropriation of different parts of the globe to ſome particular ſpecies of ſtone environing it, is contradicted too evidently by notorious facts to be now admitted. See 5 Sauſſ. 461.

CHAP.

CHAP. I.

OF PRIMITIVE MOUNTAINS.

THE principal character of primeval mountains, is the abſence of all organic remains from the interior part of their maſs and the compoſition of the ſtones and rocks of which they conſiſt. I ſay from the *compoſition**, becauſe between theſe rocks and in their veins and cavities ſuch remains are ſometimes, but very rarely, found accidentally depoſited through rifts poſteriorly choaked up, and often on their ſummits, being left there by the deluge. Pallas Loc. Cit. 44. Prince Le Gallitz. Mineralogy, p. 27.

2° They commonly form the higheſt ridges in any chain, and the moſt extenſive: this Pallas atteſts with reſpect to the Uralian and Altaiſhan chains, Born of the Tranſylvanian and Hungarian, Tilas and Bergman of the Swediſh, Haller and Sauſ-

* Haidinger 7. Mem. Par. 1747, 1072.

ſure of the Swiſs, Charpentier of the Saxon, Ferber of the Bohemian, La Peyrouſe of the Pyrenees; but frequently alſo when intermixed with ſecondary mountains their height does not exceed 2 or 300 feet.

3° They never cover ſecondary mountains, but are often covered by them. Mem. Par. 1747, 1082. Nor do they lean on the ſecondary, but the ſecondary often lean upon them and cover their flanks and inveſt them; but they often cover each other.

4° They are ſometimes ſtratified, but more frequently in huge blocks: their ſtrata never alternate with ſecondary ſtrata. Some are *unigenous*, conſiſting for the greater part, at leaſt, of one ſpecies of ſtone or aggregate; ſome *polygenous*, conſiſting of various ſpecies, alternating with, or paſſing into, or mixed with each other.

The materials of which they conſiſt, or which they contain in different inſtances, are granites and ſtones of the granitic claſſes, as granitines, granitells, granilites, ſienite, grunſtein, or gneiſs, ſhiſtoſe mica, ſiliceous ſhiſtus, baſanite, hornſlate, ſhiſtoſe or horn porphyry,

porphyry, jaſper, petroſilex, quartz, pitch-ſtone, hornblende, hornblende ſlate, argillite, trap, wacken, mandelſtein, porphyry, ſerpentine, pott ſtone, ſand ſtone, breccias, pudding ſtone, rubble ſtone, granular lime-ſtone, fluors, gypſum, topaz rock; for ſand ſtone and rubble ſtone may be primeval, being formed after the cryſtallization of the greater maſſes.

Some of theſe are common both to primeval and ſecondary mountains, as trap, argillite, porphyry, ſandſtone, breccias; paraſitic ſtones are omitted.

Among primeval mountains a diſtinction may be eſtabliſhed betwixt thoſe whoſe exiſtence preceded that of fixed air, and thoſe of poſterior formation; though in ſome parts of the globe mountains of the moſt ancient denomination, as thoſe of granite and gneiſs, ſeem contemporaneous with the exiſtence of fixed air, becauſe mountains of the ſame denomination could not all have been exactly coeval; thoſe, however, whoſe exiſtence was ſubſequent to that of fixed air, are very rare. Among primeval ſtones we may alſo diſtinguiſh the *original,* as granite, gneiſs, &c. and the *derivative,* as *rubble ſtone, breccias,* &c.

§ 1.

Of Granite, and Granitic Compounds.

By granitic compounds, I mean granitines, granitells, and granilites. As they are frequently found in the ſame mountain; among granitines and granilites, I comprehend *ſienite,* according to the proportion of its ingredients; and among granitells I reckon *grunſtein.*

Granite is moſt commonly found in huge blocks, ſeparated from each other by riſts irregularly diſpoſed; but it has alſo been found forming ſtrata, either vertical or nearly ſo, or horizontal, in Siberia. Renov. 37. Ferb. Act. Petropol. 1782, 2 Part, p. 201. 4 N. Act. 285. 1 Berg. Jour. 1791, 85, Charp. 389, and in the Pyrenees, La Peyrouſe traité des Mines de Fer, p. 329, and Deſcript. des Pyrenees, 172; and by Sauſſ. 6 Sauſſ. 317. 322: ſometimes it is found in rounded blocks with earth between them, La Peyrouſe Loc. Cit. 332.

The largeſt maſs of granite is that called the Pearl diamond, thirty miles from the Cape of Good Hope; its circumference is half a mile, and its height about 400 feet. Phil. Tranſ. 1778, 102.

In Dauoria it is frequently ſhot through or impregnated with calcareous particles, or contains primitive limeſtone as a conſtituent part, or calcareous ſpar; per Patrin, 38 Roz. 231. 235. 7 Sauſſ. 83. It is alſo frequently ſhot through with calces of iron; ſometimes it contains red or brown hæmatites, or compact brown iron ſtone, or tin ſtone, as in Leſke, G. 62, &c. Sometimes galena, or native ſilver, or black cobalt ore. 2 Widenm. 1004; gold, copper, biſmuth, martial and arſenical pyrites, and molybdena more rarely. 7 Sauſſ. 274. 2 Lenz. 335. Flints have alſo been found in neſts in it in Bohemia. Werner's Cronſtedt. 138. And various paraſitic ſtones; ſee 2 Lenz. 334. and Garnets, 3 N. Nord. Beytr. 175. Layers of limeſtone are never found in it, but lumps of gneiſs ſometimes are. Werner Kurze Claſſif. 9. Argillite has alſo been found ſtuck in granite. Born *Hungary*, 207. A lump of cryſtallized cal-

 careous

carcous ſpar has been found in the midſt of a block of granite, not occupying a particular ſpace, but entering into the compoſition of a particular part of the granite. 39 Roz. 9. And at Gedre and Gavernie, it has been obſerved incloſed between ſtrata of primitive limeſtone. 13 An. Chy. 166.

Granite, with ſienites and grunſtein, is found on the ſummit of Mont Blanc, 7 Sauſſ. 280. 288. Nay the whole mountain conſiſts of it. Ibid. 385. The higheſt mountains are of the granitic kind, and from their partial decompoſition afford lofty ſpires and various groteſque figures; they are commonly covered with moſs. Pallas 1 Act. Petropol. 24, 25. 1 Sauſſ. § 131. It is one of the rocks moſt univerſally diffuſed through the globe. Voight Prack. 29. Laſius thinks it the baſis of all the mountains in the Hartz. Laſius, 65. Maſſes of granite are often found at a diſtance from the mountains to which they belong; theſe, incredible as it may ſeem, have, in ſome caſes, been carried off by an immenſe torrent. Thus in 1775, a ſudden ſoutherly wind having partly diſſolved, and partly looſened, an enormous maſs

maſs of ice, it was carried down to a narrow paſſage, which ſtopped for ſome time the waters that uſually flowed down through that paſſage; but theſe at laſt collecting, forced it down that paſſage, and daſhed it with ſuch inſuperable violence againſt the obſtructing rocks, as to bear down in its deſcent, to conſiderable diſtances, granitic maſſes of ſeveral hundred cubic feet. 1 Helvet. 9, 10. In ſome caſes, where higher mountains of another claſs intervene, theſe blocks might have rolled down, or have been carried down, from granitic mountains, originally much higher, but ſince degraded, and before the vallies that now ſeparate them from the intervening mountains were formed. Such as are found on the ſummits of mountains, might have been the remains of the decompoſition of ſtill larger and higher maſſes; or may have been ſhattered by lightning. Some may have been bolted off by the ſhock of an earthquake; ſome may have been fragments of a granitic mountain ſwallowed up, after various convulſions, by the earth (as near Geneva; ſee 33 Roz. 8.); and ſome may be regenerated granite, reſulting

ſulting from the reunion of the ſand of a decompoſed granite. Granite and ſienite have, in ſome inſtances, been found ſuperimpoſed on other rocks of the primeval claſs. Pallas found granite repoſing on *argillite*; and ſienite, on *argillite* or *gneiſs*. 3 Helvet. Magaz. 175. 2 Pallas Reiſs, 517. 520. And alſo Soulavie in the Cevennes, 3 Soulav. 162. And La Peyrouſe in the Pyrenees, Sur les Mines de Fer, 329. Or on *ſerpentine*, or a compound of ſerpentine and Iade. Ibid. Voight on *hornblende ſlate* at Ehrenberg. 2 Berg. Jour. 1790, 300. Soulavie on primeval limeſtone; and not piercing through it, as ſome have imagined, for a gallery was worked through the limeſtone under the granite, without meeting the granite. 1 Soulavie France Meridionale, 374, 375. 377, 378; and the anonymous author of Deſcription des Pyrenees, 144. But maſſes of granite have often *tumbled on limeſtone*, and hence have been erroneouſly thought to have originally repoſed on it, per Ferber, Act. Petrop. 1783, 298, 299. But far oftner, and almoſt univerſally, granite ſerves as a baſis on which other rocks, both primitive and ſecondary, reſt. Thus, it it underlays *argillites*, *gneiſs*, and *limeſtone*, moſt commonly, per Ferber, Act. Petrop. 1782, 2 part, 208. *granular limeſtone*, *jaſper*, *breccias*, *ſerpentine*, and *porphyry*. 1 Gerh. Geſch. 66. 3. Soulavie, 72. Ferber *Bohemia*, 106. And in the Alps, *gneiſs*, *ſhiſtoſe mica*, *argillite*, *hornſlate*, and *limeſtone*, repoſe on *ſienite*. 3 Helvet. Mag. 175. 4 Helvet. Mag. 267. 312. 315. The mountain of Taberg in Sweden, reſts on grunſtein, conſiſting of felſpar and hornblende. 2 Bergm. Jour. 1789, 2002. It is alſo frequently ſurrounded and inveſted with rocks of various ſpecies, which lean on it at various heights, but which its ſummit ſurpaſſes. Thus in the Tyrole, the granite pierces through, and riſes above, the *argillite* that ſurrounds it. Ferber Italy, 46. And in the Pyrenees. 13 An. Chy. 164. It ſometimes alternates, and ſometimes is mixed with other ſtones, ſee § 20. It ſometimes decompoſes into concentric layers. Charp. 31.

Granite is alſo ſometimes newly formed, not as to its individual component particles, but in reſpect to the reunion of the grains that compoſe it. A ſtratum of this ſort, Mr.

Mr. Gerhard met with at Schreiberhau in Silesia, and D'Arcet near Bareges, over argillite and limestone. 1 Ger. Gesch. 68. This regeneration takes place, as Lasius well remarks, when the granitic sand lies in a damp situation, and screened from a free access of air; thus he found granitic sand, employed in filling a dyke in the Elbe, to have hardened to such a degree, in the space of sixty years, as not only to prevent the passage of water, but to present solid masses, scarcely distinguishable from native original granite, and, where contiguous to this last, to be difficultly separated from it by a blow. Lasius *Hartz*, 91, 92. Friesbben, 189. Mr. De La Coudroniere also observed on the banks of the Mississippi, mountains of sand half converted into granite, and in the midst of one of them a branch of green oak petrified. 21 Roz. 237. Granite is also found in veins, and consequently of modern formation, as Werner observes. Werner Enstehung der Gange, § 49. and 1 Sauss. § 600, 601. Bartollozzi found a passage anciently wrought to extract crystals in mount Baveno, filled by a vein of regenerated granite, 21 Roz. 468.

Lavoisier

Lavoiſier alſo found near the mountain of St. Hypolite, two alternating beds of indurated granite, or, as he calls it, *true granite*, intercepted betwixt two beds of granitic ſand that lay over each: this muſt have been regenerated granite, formed by infiltration from the ſuperior granitic ſand. Mem. Par, 1778, 439. Sauſſure found it alſo in the fiſſures of gneiſs. 1 Sauſſ. 601. Pallas thinks that all ſtratified granite is formed of the decompoſed grains of the primitive blocks. 2 Nev. Nord. Beytr. 366.

§ 2.

Of Gneiſs, and Shiſtoſe Mica.

I treat of theſe together, as they are chiefly diſtinguiſhed from each other by the abſence, or preſence, of felſpar; in moſt other particulars they agree, inſomuch, that ſome authors (1 Gerhard Geſch. § 62.) comprehend both under the name of *gneiſs*. *Stellſtein* alſo is another name by which it is often denoted. Gerh. Ibid. and Berg. Kalend. Both are ſtratified. 1 Ger. Geſch. § 51, and generally in the direction of the mica. Voight Prack. 32.

Gneiſs

Gneiſs frequently contains huge maſſes or layers of *granular limeſtone*, as Sauſſure obſerved in Mount Simplon. 37 Roz. 7. And Charpentier in Saxony. Charp. 173, 174. And blocks of *granite* perfectly incorporate with, and paſs into it. Charp. 391. 2 Sauſſ. § 676, and 8 Sauſſ. 64. And huge maſſes of *felſpar*. 1 Gerh. Geſch. 85. Or vaſt layers of *porphyry*, as at the mountain of Kimerſdorf, either argillaceous, or petroſiliceous. 2 Berg. Jour. 1790, 455, 456, 457. Often alſo ſtrata of *hornblende ſlate*; and *hornblende* ſometimes enters into its compoſition. Berg. Kal. 202. Trap alſo has been found in it. 4 Helv. Mag. 546. It is alſo remarkably metalliferous; moſt of the Saxon and Bohemian mines are ſeated in it. Berg. Kal. 203. Shiſtoſe mica alſo contains, ſometimes, beds of primeval *limeſtone*, or *hornblende ſlate*. Berg. Kal. 204. Alſo calcareous ſpar, ſappare, and garnets, as already mentioned. In metallic ſubſtances it is nearly as rich as gneiſs. Mountains of gneiſs are not ſo high nor ſo ſteep as thoſe of granite, Mount Roſe in Italy, and a few others, excepted; and their ſummits are generally more rounded.

ed. Gneiſs reſts moſtly on *granite,* but ſometimes on *argillite, porphyry, ſerpentine, granular limeſtone,* or trap. Per Charpentier. 4 Helvet. Mag. 545, 546. 2 Lenz. 341. Or ſandſtone. 2 Lenz. Ibid.

It ſometimes underlays *granite* (ſee granite). At Montevideo in Mexico Granite and gneiſs alternate with each other. 1 Berg. Jour. 1789, 193. It ſometimes alternates with argillite. 5 Sauſſ. § 1219. Sometimes with granular limeſtone. Charp. 2 Lettere Oritologice al Signor Arduino. Shiſtoſe mica alſo reſts on granite, or porphyry. 2 Lenz. 345. It underlays argillite, limeſtone, and ſandſtone. 2 Lenz. 345. Voight Prack. 38.

Where gneiſs is contiguous to granite, its quartz and felſpar are more apparent, and the micaceous part leſs predominant. 1 Gerh. Geſch. 78. And where moſt diſtant from granite, the contrary happens, and it often graduates into argillite. Ibid. Or into ſiliceous ſhiſtus. Id. 83. The nearer it is to metallic veins, the more earthy or ſofter it becomes. Id. 80. Charp. 79. That gneiſs and granite were frequently at leaſt contemporaneous, appears from this, that metallic

metallic veins run without interruption from one to the other. Charp. 256. Sometimes a disintegrated granite reappears in the form of gneiss. Flurl *Bavaria*, 3:0.

§ 3.

Siliceous Shistus, and Basanite.

In upper Lusatia an entire mountain is formed of it. Karsten on Leske, G. p. 21. Charp. 24. And no petrifactions are found in it. Ibid. 26. And in Siberia, per Herman, in 1 Berg. Jour. 1791, 82, and as Renov. 31. says of blue hornslate, which I suppose to be either siliceous shistus, or basanite. It is also found in the Alps between gneiss and hornstone. 4 Helv. Mag. 115. Yet Werner in Pabst's catalogue, p. 236, places it among secondary rocks; Voight places it among the primary (but confounds it with siliciferous argillite. Prack. 43), and Charpentier, 4 Helv. Mag. 547. Schlangenberg in Siberia mostly consists of it, mixed with hornblende and felspar. Renov. 86 and 89. It is found among the primitive rocks of Altai. Renov. 86. He calls it hornshiefer. It frequently occurs in argillitic mountains.

mountains. 2 Lenz. 351. It often makes right angles with the argillite, and passes into it. 3 Helv. Mag. 252. It frequently forms high grotesque cliffs, *reposes* on, and even strongly adheres to granite. 2 Friesleben, 203. Charp. 22. It seems to be what Lasius calls *filicited trap*. N. 73. whose specific gravity is 2,685. Lasius, 124. When black it seems intermediate between siliceous shistus and Lydian and quartz. Friesleben, 201. Others also call it *hornfels*. When mixed with hornblende it looks greenish grey, and becomes heavier. Friesleben, 208. May it not be the petrosilex, 2 Sauss. 4°. § 1045. and which he call *palaiopetre*. Vol. 5, in 8vo. § 1194? The only difference is, that its transparency seems greater than that of siliceous shistus. The thinness of the lamellæ of Saussure's seems to proceed from horizontal cracks. See also 5 Sauss. in 8vo. p. 77, § 1223.

§ 4.

Jasper.

Mountains of striped jasper occur in Siberia, and often with breccias, but without petrifactions, per Herman. 1 Berg. Jour. 1791, p. 84 and 94. Of red jaſper, Ibid. 88. And alſo of green jaſper. 2 Gmelin, 81. (French.) It often forms thick ſtrata in mountains of ſhiſtoſe mica in the Appenines. Ferber Italy, 109. and in Siberia, 2 Herm. 281. In Saxony it is found alternating with, and ſometimes mixed with, compact red iron ſtone. 2 Berg. Jour. 1788, 485.

In the ſouth of France it occurs repoſing on granite, and underlaying baſalt. 3 Soulavie, 72. In the Altaiſchan mountains it has never been found in contact with granite, but it ſometimes underlays argillite. 6 Nev. Nord. Beytr. 115.

§ 5.

Hornſtone, Petroſilex.

This is one of thoſe ſtones which occurs of primary, as well as of ſecondary formation. In Siberia, mountains of hornſtone in which fragments of hornblende and felſpar are diſperſed, occur. Renovantz, 31. And hornſtone penetrated with limeſtone, is found in mighty layers in Douria, but without

without petrifactions. 1 Chy. Ann. 1791. 155 and 345. And Patrin in 38 Roz. According to Dolomieu, 44 Roz. 247, petrosilex is found only in primitive mountains; but the contrary will be shewn in Chap. II. § 4. At Menard in Forez, it is frequently found mixed with pitchstone, per Bowman. 30 Roz. 377. But query, May we not distinguish with Saussure a primary and secondary? Near Bidart black hornstone is found alternating in thin layers with a coarse (seemingly primary) limestone. Descript. Pyren. 2. Saussure found it on the summit of Mont Blanc. 7 Sauss. 275. 287.

§ 6.

Quartz.

The mountain of Kultuck, on the S.W. end of the lake Baikal, 350 feet high, and 4800 long, and still broader, consists entirely of milk-white quartz; per Laxman 1 Chy. An. 1785, 265. Also Flinzberg in Lusatia, almost entirely. 2 Berg. Jour. 1789, 1054. There is also an extensive narrow ridge of quartz, some miles long, in Bavaria. 2 Berg. Jour. 1790, 529, &c.

Flurl Bavaria, 309. Monnet mentions a rock of quartz 60 feet high. 17 Roz. 163. Mountains of it alſo occur in Thuringia. Voight Prack. 69. and in Sileſia. Gerh. Beytr. 87. and in Saxony. 1 Berg. Jour. 1788, 269. and in layers between gneiſs and ſlate mica. 2 Lenz. Alſo in Scotland. 2 Wms. 52. It is not metalliferous. Werner Kurze Claſſif. 15. Petrol is often found in it. 1 Berg. Jour. 1791, 91. The mountain of Swetlaia-Gora, among the Uralian, conſiſts of round grains of quartz, white and tranſparent, and of the ſize of a pea, united without any cement. 2 Herm. 278.

§ 7.

Pitchſtone.

In Miſnia, it forms entire mountains. 2 Berg. Jour. 1788, 491. And in other mountains huge ſtrata that alternate with porphyry, and as they contain abundance of quartz and felſpar may be called pitchſtone porphyry. Ibid. and 1 Emerling, 264.

I do not find it mentioned among primitive ſtones, except when a porphyry; ſee Karſten

Karſten in Leſke. It is often the ſubſtance of which petrified wood conſiſts. It ſometimes alternates with granite. Charp. 63.

§ 8.

Hornblende and Hornblende Slate.

Mountains of black hornblende exiſt in Siberia. Revantz. 32. as the Tigereck. 4 Nev. Nord. Beytr. 192. and others mentioned by 2 Herm. 271. Frequently mixed with quartz, mica, or felſpar, or ſhorl, and either greeniſh or black. Ibid. But it is more commonly found in mighty ſtrata, as in Saxony; or ſtill oftener as a conſtituent part of other primeval rocks, as in ſienite and grunſtein; ſometimes in layers in gneiſs, or granular limeſtone, or argillite; and ſometimes in horn porphyry. 2 Berg. Jour. 1788. 508. 1 Lenz. 325. 1 Emerling, 325. or in the gullies of granite. Herm. Ibid. hornblende ſlate was obſerved among the primemeval rocks on the aſcent of Mont Blanc. 7 Sauſſ. 241. 253. mixed with plombago. Ibid. and on its ſummit. Ibid. 289.

Strata of ſhiſtoſe hornblende occur ſometimes in gneiſs, as already mentioned. At

Miltiz a ſtratum of it has been found over granular limeſtone. Voight Prack. 33. In lower Sileſia it has been found on ſienite. 4 Berl. Beob. 349. Granite ſometimes reſts on it. 2 Berg. Jour. 1790, 300. Voight Mineral. Abhandl. 25. Hence there can be no doubt of its being a primitive ſtone. A mountain of it exiſts in Tranſylvania. 1 Bergb. 40. Nay, granite has been found in it. 1 Berg. Jour. 1789, 171. It is frequently mixed with mica, more rarely with viſible quartz. Emerling.

§ 9.

Indurated Lithomarga.

This often occurs in the rifts, or joints, or intervals, of beds of gneiſs, porphyry, ſerpentine, and alſo in topaz rock, and, therefore, muſt be primitive, if not deriving from decompoſition. It often alſo forms neſts in traps, baſalts, or amygdoloids. Emerl. 359.

§ 10.

Argillite.

It forms whole mountains, Voight Prack, 38. But more commonly, only partially enters into them, as in Saxony. Charp, 175. Or entire ſtrata, as at Zillerthal in Tyrole, its mountains are of gentle aſcent.

There is no doubt of its being often primitive, for in Saxony it frequently alternates with gneiſs and ſhiſtoſe mica. 3 Helvet. Mag. 190. 1 Berg. Jour. 1792, 536. And with primitive limeſtone. 8 Sauſſ. 144. And in Hanover granular limeſtone is found betwixt its layers. 1 Bergm. Jour. 1791, 306. We have alſo ſeen that both granite and gneiſs often reſt upon it. Both Karſten 3 Helvet. Mag. and Monnet in 25 Roz. 85, ſufficiently eſtabliſh this diſtinction. There are two ſorts of it particularly to be attended to, the *harder* and the *ſofter*; the *harder* border upon and often paſs into ſiliceous ſhiſtus, or baſanite, or hornblende ſlate. The *ſofter* border upon or paſs into *trap*, or wacken, or rubble

 ſtone,

ſtone, or rubble ſlate, or coticular ſlate, or indurated clay, and the harder often graduate into the ſofter. 3 Nev. Nord. Beytr. 169. Or border upon the *muriatic genus* and paſs into ſhiſtoſe chlorite, or ſhiſtoſe talc, or gneiſs or ſhiſtoſe mica. It often contains quartz, both in veins and betwixt its laminæ. Voight Prack. 41. More rarely felſpar, ſhorl, garnets or hornblende and granular limeſtone. Berg. Kal. 205, 206. The ſofter ſorts are remarkably metalliferous. Bergm. Kal. Voight Prack. 40. The famous mountains of Potoſi conſiſt of it chiefly. 1 Bergm. Jour. 1792, 545. In Saxony it is found in primitive limeſtone, 2 Bergm. Jour. 1792, 134, and often mixed with it, as in Leſke, G. 328. It is ſo much the more ſiliciferous as it approaches more to granitic mountains. Laſius 121. It paſſes into rubble ſtone. 2 Bergm. Jour. 1788, 498. In the argillites of the Pyrenees no organic remains are to be found. Deſcrip. Pyren. 27. Sauſſure found it in the ſnowy regions of Mount Blanc. 7 Sauſſ. 256.

§ 11.

§ 11.

Trap.

On the ſubject of this rock I muſt enter into a long detail, on account of the various falſe or imperfect deſcriptions given of it, and the various ſpecies of ſtones that have, on that account, been erroneouſly denoted by this appellation.

In the firſt place it muſt be allowed that as this name is come to us from Sweden, the deſcription and properties of it given by the moſt exact Swediſh Mineralogiſts ſhould be thoſe which we ſhould alſo adopt and denote by that *name*. Among thoſe Mineralogiſts I ſelect Cronſtedt, Bergman, Wallerius and Hermelin. Cronſtedt defines *trap* to be a rock formed of ſoft martial jaſper, or an indurated martial clay. § 265 of Brunich's, or 267 of Engeſtrom's Engliſh tranſlation. " Its *colour* is dark grey, as " that of the top of Kinneculla, and that " of Kunneberg in Weſt Gothland, or " *black*, as that of Stahlberget in Dalarne, " or Hallefors, Salbergmine, Norberg in " Weſtmanland and Oſterſilverberget in " Dalarne; or *grey*, as that of Dalwick, or " *bluiſh*,

" *bluiſh*, as that of Oſterſilverberget, or *deep* " *brown*, as that of Gello in Norway, or " *reddiſh*, as that of Bragnas in Norway, or " Dalſtugun in Dalarne. It forms whole " mountains, as Hunneberg, or Drammen, " but is oftener found in the form of veins " in mountains of another kind, running in " a ſerpentine manner acroſs the direction " of the rock. Where it is preſſed cloſe " it ſeems to be perfectly free from he- " terogenous ſubſtances. But where not preſ- " ſed, it appears leſs homogeneous. When " very coarſe, it is interſperſed with *felſpar*, " but it is not certain that the finer ſorts " contain any. Beſides there are ſome " fibrous particles in it, and ſomething that " reſembles calcareous ſpar, but which does " not effervеſce with acids, and melts as " eaſily as the ſtone itſelf into a black ſolid " glaſs" We have of that found on the ſummit of Kinneculla, (which conſequently is leaſt compreſſed and apparently contains heterogenous particles) 2 ſpecimens in Leſke's cabinet G. 252. and S. 145. The colour is grey, as Cronſtedt deſcribed, from a viſible mixture of black hornblende and white quartz; but the hornblende in

far

ſar greater proportion; its luſtre 2, or 0; and that of the quartz 1, or 0; a few particles of yellowiſh felſpar may alſo be diſtinguiſhed in both. Cronſtedt diſtinguiſhes three varieties, namely, the coarſe grained ſpicular, the coarſe grained granular, and the fine grained: theſe laſt he calls *touch ſtone,* which are thoſe already mentioned of Hellefors, Norberg, Dalwich, and Dolſtugun. The fracture of the firſt, which are from the ſummit of Kinneculla, I found to be the *coarſe grained earthy*; hardneſs 9; ſpecific gravity of one ſpecimen 2,949, of another 2,947. " Calcined it cracks and " reddens, and on eſſaying it, yields 12 or " more per cent. of iron; no other ſort of " ore is found in it, unleſs now and then " ſomewhat ſuperficial in its fiſſures, for it " is commonly cracked, even to a great " depth, in acute angles, or in the form " of large rhomboidal dice. By expo- " ſure to the air it ſlightly decays, and " leaves a brown powder." The fine grained black, which Cronſtedt called touchſtone, is now allowed to be that now called baſalt, and, conſequently, differing from trap only in grain, and in being, apparently, more homogeneous; fee Karften on 1 Lefke, 201. But its compofition is evidently the fame, as will appear, ftill more clearly, from Bergman's account.

Bergman in his letter to Vantroil, p. 392, tells us, " that trap is generally found in " fquare irregular cubes, whence it pro- " bably obtained its denomination on ac- " count of fome fimilarity with ftones ufed " for ftair-cafes. It is alfo found in prif- " matic triangular forms, though rarely; " as alfo in the form of immenfe pillars, " as thofe called *traelftenar* at the foot of " Hunneberg, which *have feparated them- " felves* from the remaining part of the bed." And in his account of the mountains of Weft Gothland, he fays, that in the fpring feafon, this ftone often cracks into rhombic fragments. 5 Berg. 116, 117. On comparing the bafaltic pillars of Staffa with the fine grained traps, he found their colour, fracture, and hardnefs, the fame; the fpecific gravity of the Staffa pillars was 3,000, that of the trap 2,990. Their relation to fire, and to acids, the nature and proportion of their component parts, he found alfo to agree as nearly as poffible.

Voutroil,

Voutroil, 394, &c. 3 Bergm. 212. He tells us, that in Sweden it forms slender veins in primeval mountains. Ibid. 214. Wallerius also adheres to Cronstedt's definition. He calls it *corneous trapezius*. Sp. 172. Its colour, he tells us, is black, grey, bluish, greenish, or reddish; the black forms a sort of touchstone, and takes a polish,—occurs in mountains divided into cubes or rhomboidal figures; fracture conchoidal, sometimes presenting scaly, sometimes spicular, concretions; hardness 9, specific gravity 2,800; heated quickly, it decrepitates; if more slowly, it hardens, and at last melts into a solid black glass. It sometimes slightly effervesces with acids. It contains from 8 to 15 per cent. of iron. He refers it to the same mountains as Cronstedt. Baron Hermelin also calls the upper stratum of Kinneculla, a trap, and refers to Cronstedt; he allows it from 8 to 16 per cent. of iron. Schwed. Abhandl. 1767, 26. but he means iron in its metallic state. Hence we may infer, with Bergman, that the close grained trap and basalt are exactly the same, both as to external and internal characters, and that the fine grained differs

from

from the close grained, only in the minuteness of the particles that compose them, arising solely from compressions as Cronstedt well remarked. The quantity of iron, we see, is variable, from 8 to 16 per cent. in the dry way; but in the moist way, Bergman found it to amount to 25 per cent. This, however, I believe is rather too much; I believe it rather to be from 12 to 22 per cent. and hence from the difference of a finer or coarser grain, the variation of the specific gravity from 2,800, or at least from 2,78 to 3,000, or 3,021. The specific gravity of *ferrilite* I marked too low, for I find it 2,800.

Many are inclined to place trap among *aggregates*, or compound stones, on account of the number of foreign stones that are often found stuck in it, and which I have already enumerated in my first volume; but as this composition is neither determinate nor constant as it is in granites and porphyries, I think it proper to leave it among the simple stones, all of which are susceptible of such compound appearances; notwithstanding the easy fusibility of trap and basalt in clay crucibles, Klaproth found

found it infuſible in a crucible made of charcoal. 1 Klapr. 7. Probably becauſe the iron, to which it owes its fuſibility, had abſorbed too great a quantity of carbon, and was thus converted into plombago; or, rather, becauſe the iron to which its fuſibility is due, was reduced to a metallic ſtate. Gerhard tells us, that a ſtone, which he calls *jaſpis trapezius*, and which, he ſays, is the trap, and *corneus trapezius* of Wallerius, was not melted in a clay crucible. 2 Gerh. Geſch. 22. N. 52. poſſibly it may have been a real jaſper that he uſed, or perhaps a trap ſo highly impregnated with carbon as to be difficultly fuſible.

Renovantz tells us, that the Ruſſian government of Olonitz abounds in mountains of trap, ſeveral ſpecimens of which occur in Leſke *S.* from N. 90 to N. 109; but all of them are not genuine trap, for inſtance. N. 94, of which I found the ſpecific gravity to be only 2,747; and as Karſten remarks only a few ſpecks of acicular hornblende are diſcernible in it; its colour is dark greeniſh grey, from a mixture of minute white ſpecks, and in ſome places browniſh grey; its fracture uneven or earthy;

earthy; hardneſs 9, it therefore contains a mixture of *wacken*.

Sienite is alſo often miſtaken for trap; thus the ſpecimen in Leſke, *S*. 1316, approaches nearly to trap, but its ſpecific gravity is only 2,713; its colour is *dark grey*, from a mixture of white particles of various ſizes, with a ground which at firſt ſight appears *black*, but on cloſe inſpection, or viewed through a lens, appears *reddiſh black*. Luſtre 0; tranſparent where thin, 1 or 2; the tranſmitted light is whitiſh and yellowiſh; fracture coarſe grained earthy; hardneſs 9. Again, the ſpecimen *S* 504; Karſten calls a "ſtone in which a large pro-"portion of quartz and hornblende are "mixed, and perhaps a *variety of trap*." This is N. 4 of Voight's firſt collection; and in the catalogue annexed to his Geological Letters, p. 68, he calls it a "gra-"nitic mixture of hornblende and a whitiſh grey quartzy ſubſtance." Yet he does not allow it to be a quartz, becauſe it difficultly gives fire with ſteel. Afterwards in his Mineralogical Deſcription of Ilmenau, he calls it a "granitic mixture of horn-"blende, quartz, and felſpar," p. 8. In

the

the next page he calls it a *grunſtein*, as hornblende is the principal ingredient. In his ſecond catalogue, p. 13, N. 5, he calls it a granitic mixture of black hornblende and white compact felſpar, and thinks it may be called a *ſienite*; its colour is ſpeckled grey, from a viſible mixture of black hornblende and a greyiſh white ſtone, which has, indeed, a quartzy appearance, with nearly an equal proportion of hornblende and ſome dark red knobs, that appear to be dull garnets. Luſtre of the hornblende, in a direct view 2, in an oblique 0; of the white ſtone 1 or 0; tranſparency of the hornblende 0, of the white ſtone 1, tranſmitting a yellowiſh light; fracture, coarſe grained *earthy*, tending to the *uneven*; hardneſs 9; ſpecific gravity 2,933; hence it appears to be a trap; and the white ſtone cannot be a common felſpar, elſe its ſpecific gravity would be lower. Laſius thinks that trap and baſalt can be diſtinguiſhed only by their ſituation in their natural *abodes*. (*Lagerſtadt.*) Baſalt, he ſays, never has any regular direction, whereas trap has both *direction* and *dip*. Hartz. 128. Yet Karſten, near Winneburgh, found baſalt in ſtrata that dipped to the E. under an angle of 60°, ſtretching from N. to S. 1 Bergm. Jour. 1788, 337. Werner cites other inſtances in 2 Bergm. Jour. 1788, p. 890 and the annexed note. Charpentier alſo mentions layers of it, p. 19. Widenman comprehends both trap and wacken under the ſame ſpecies, though they differ much in ſpecific gravity, and hardneſs, and grain; and wacken never contains compact felſpar, as trap often does: Charpentier diſtinguiſhes them exactly, though he thinks the Swedes did not. *Saxony*, p. 187. Hence he conſtantly denotes both trap and baſalt by the name *baſalt*. According to Werner, baſalt, wacken, hornblende, and ſhiſtoſe porphyry, are of the ſame formation; the Swediſh mineralogiſts, he thinks, comprehended them all under the name of *trap*. 2 Berg. Jour. 1789. Yet it is evident that neither Cronſtedt, Wallerius, nor Bergman, confounded hornblende and trap: as to baſaltic pillars (except thoſe of Hunneberg) they appear not to have been acquainted with them; Bergman indeed knew them, and diſtinguiſhed them only as varieties, as I do.

Karſten, in his catalogue of the Leſkean cabinet, denotes both trap and baſalt by the name of *baſalt*, except in the geographical part; he thinks trap ſhould be reckoned an aggregate of hornblende, felſpar, and quartz, the quartz only viſible in white ſpecks, and the felſpar in oblong particles, imbedded as in ſhiſtoſe porphyry, and the hornblende the prevailing ingredient. Note to Leſke, *G* 269. Specular and magnetic iron ſtone, and calcareous ſpar, he thinks peculiar to it, as garnets to ſlate mica; Charpentier and Noſe alſo conſtantly denote what I call trap, by the name of baſalt. Dolomieu in 44 Roz. 256, deſcribes trap by contraſting its characters with thoſe of a ſtone, which he calls *roche de corne*; a ſtone which he diſtinguiſhes from hornblende, and whoſe characters are deſcribed only by contraſting them with thoſe of *trap*. As I can form no idea of this ſtone, I confeſs the whole to me is unintelligible. Faujas in his treatiſe on traps, p. 2, juſtly remarks that " the denomination *roche de* " *corne* is purely ideal, and only ſerves to " lead into error." The Germans avow they do not know how to tranſlate it. I

Berg. Journ. 1792, 459, and 2 Berg. Journ. 1792, 468, in note. The learned anonymous author of a deſcription of the Pyrenees, ſays, " this denomination has been applied to ſo many different ſubſtances, that I dare not uſe it." Pref. XV.

Faujas, deriving his knowledge from the Swedes, has given the beſt account of traps hitherto extant; it is only to be regretted that he has omitted its external characters: he allows it can be diſtinguiſhed from baſalt only by local circumſtances. The name baſalt he confines to this ſubſtance when of volcanic origin; an origination to which it is well known he is ſtrongly attached, and which I have endeavoured to refute in the firſt volume of my Mineralogy: he does not diſtinguiſh it from wacken. This ſeems, however, to be the ſixth variety of his homogeneous traps.

Karſten thinks trap compounded of hornblende, quartz, and felſpar, and certainly the ingredients of theſe are found in it, but not in the ſame proportion as in any mixture of the three, as I have already ſhewn: he does not extend this compoſition to baſalt; yet, whatever are the ingredients of trap, the ſame

ſame muſt be thoſe of baſalt, as they differ only in *ſolidity and figure*, and the ſolidity ariſes entirely from compreſſion, as Cronſtedt has remarked; the various cryſtallized ſtones that may occur in trap, are merely paraſitic, and baſalt is not deſtitute of them, only they are ſmaller.

To form an idea of the compoſition of this ſtone and the adjacent ſpecies, we muſt conſider the educts of its analyſis, by Bergman, Gerhard, Faujas, and Sauſſure: theſe are,

Baſalt and compact trap.	Baſalt.	Baſalt of Staffa.	Baſalt of Antrim.	Baſalt of Chenevari in the Vivarois.
Bergman *.	Gerhard †	Faujas ‡.	Faujas.	Faujas.
Silex.....50	60	40	46	40
Argil 15	10	20	16	20
Calx..... 8	8	12	10	8
Magneſia . 2	0	5	3	6
Iron 25	22	21	22	24

* 3 Bergm. † Crell Beytr. 3 Stucke, p. 5.

‡ Faujas Sur les Traps, 66, &c.

A Trap, or *Pierre de Carne Dure.*	A decayed Basalt.	Prismatic porous Lava from Iceland.
Saussure *.	Gerhard †	Gerhard ‡
Silex... 51	33	62
Argil .. 16,6	35,5	20
Calx... 8,4	2,5	2
Magnes. 3	—	0
Iron... 12	29	16
This is evidently imperfect.	of yellow calx.	

Hence, taking Bergman's analysis for a standard, we learn, 1°, That the siliceous part may be increased or diminished $\frac{1}{5}$, without any change of the species, the argillaceous $\frac{1}{3}$, the calcareous may be increased $\frac{1}{3}$, but not diminished. The magnesia may be absent, and the iron may be diminished a few grains.

2°. We learn to distinguish decayed ricketty basalts from porous lavas. In the first, we find the quantity of iron much the same as in undecayed basalts, for 29 per cent. of the yellow calx contains nearly the same quantity of metallic iron as 25 of the black calx, which is that found in basalt, but the quan-

* 2 Saussure, 135. † 1 Crell Beytr. ibid. 10, 11. ‡ Ibid. 12.

tity

tity of ſilex is diminiſhed conſiderably; the proportion of argil, on the other hand, is doubled, and that of calx reduced $\frac{3}{4}$, whereas in the lava the quantity of ſilex is increaſed, and alſo that of argil ſomewhat, but that of calx and that of iron conſiderably diminiſhed: this may go a great way towards deciding the controverſy touching the ſuſpicious foſſils of Fulda, &c.

With reſpect to the kindred ſpecies of rocks or ſtone, we may lay it down as a certain truth, that if the proportion of one or more of the conſtituent parts of baſalt or trap be notably altered (the others remaining with reſpect to each other in the original proportion, or one of them vaniſhing), either the ſpecific gravity, colour, or hardneſs, or all three, will alſo be altered; a *notable* alteration is, that which exceeds the limits already mentioned.

Thus, if the *proportion of ſilex be increaſed to its utmoſt limit nearly*, and that of iron diminiſhed about $\frac{1}{3}$, the other ingredients *conſtant*, we ſhall have *toadſtone*, and the whins whoſe ſpecific gravity exceeds or amounts to 2,780. If the quantity of iron be reduced to 10 or 12 per cent. we ſhall probably have the lighter whins, whoſe ſpecific gravity is between 2,72 and 2,78, and to theſe the name of *whin* ſhould excluſively be confined, the heavier being traps. If the quantity be reduced to 7 or 8 per cent. this will probably form the *hornſlate* of Voight, or, if it contains felſpar, the *ſhiſtoſe porphyry* of Werner*. If the proportion of ſilex be increaſed, and the argil eliminated, and the iron nearly ſo, with ſome addition of carbon, then we ſhall have *ſiliceous ſhiſtus.*

On the other hand, if the proportion of ſilex be at its *loweſt limit* nearly, and that of argil increaſed 2 or 3 fifths, and that of magneſia increaſed to 12 or 15 per cent. that of iron being conſtant, or nearly ſo, and that of calx diminiſhed a few grains, we ſhall have *hornblende.*

Again	50 Silex 30 Argil 15 Calx 5 Iron	ſhould I imagine give *wacken*, Charp. 186; this is ſometimes efferverſcent, ibid. and if the calx be eliminated, and in its place, iron ſubſtituted, *ferrilite* will reſult.

* I believe there muſt be ſome error in Weigleb's analyſis, or it was ſiliceous ſhiſtus he analyſed.

Laſtly,

Laſtly, 46 ſilex, 26 argil, 8 magneſia, 4 calx, 14 iron, gives the heavier argillites whoſe ſpecific gravity paſſes 2,800.

As in the ſeries of colours exhibited in the priſmatic *ſpectrum*, ſeven principal colours are diſtinguiſhed, notwithſtanding the numerous ſhades through which they paſs into each other; ſo I think a natural ſeries of ſtones might be formed, taking in not only the compoſition, but alſo the texture, grain, and ſpecific gravity: even imaginary compoſitions when grounded on analogy, are of uſe towards fixing our ideas until real and ſecure analyſes are obtained.

That trap is often of primeval and often of ſecondary formation, will appear by the obſervations I ſhall now mention, and thoſe that will occur in the next chapter.

1°. It is found in huge ſtrata in the midſt of gneiſs, per Charpentier, 4 Helv. Mag. 545, 546, and cryſtalliſed mica in the midſt of it. Gerhard found veins of it in the midſt of gneiſs, at Krobſdorf in Sileſia, 3 Crell. Beytr. 3 Stuck. 7. And Reuſs, in Bohemia, Mittel. Geb. 94, 196: alſo in granite, 2 Berl. Beob. 197, Ferb. 4 Helv. Mag. 156.

2°. It has been found alternating with

granite near St. Malo, Faujas *Trap* 86, and with gneiſs, Charp. 187.

3°. The nearer trap is to granite, the harder it becomes, which ſhews they originated at the ſame time, Laſius, 121. Veins of quartz about one foot deep, have been found on the ſurface of a mountain of trap, 1 Nev. Nord. Beytr. 143, 145. Mandelſtein (or Amygdaloid) ſometimes belongs to primary, ſometimes to ſecondary mountains, Werner, 13; it forms whole mountains in the territory of Deux Ponts, Laſius 259; in Norway it repoſes on granite, Haiding 52. Wacken is alſo found in gneiſs and ſlate mica, Charp. 81. 3 Helv. Mag. 236; it even alternates with gneiſs, Charp. 187.

As trap contains aerated calx, it is plain its formation was ſubſequent to that of fixed air; but ſo alſo were ſome granites and other primitive ſtones, for it is not to be imagined that all were every where formed at the ſame exact time.

Trap ſometimes alternates with argillite, as at Leidenberg, 1 Noſe 209. Sometimes with ſerpentine, 1 Nev. Nord. Beytr. 146. Black limeſtone has frequently been miſtaken for trap, Noſe Samlung 274, 278.

I am

I am convinced that ſerpentine alſo, eſpecially when containing hornblende, has been taken for trap, and vice verſa. Porphyry containing zeolyte has alſo been thought a trap, Leſke G. 237.

Primeval trap is frequently metalliferous, per Ferber, 4 N. Act. Petropol. 286; it contains iron and copper ores, Ronov. XII. 1 Nev. Nord. Beytr. 140, 143. At Scheibenberg a ſhaft has been driven under ſome baſaltic pillars that ſtood upright on the ſummit of the mountain, but nothing baſaltic was found, Charp. 224, 2 Bergm. 1. 849, 852: the pillars are alſo ſupported by horizontal ſtrata, ibid.: ſo that the idea of their being thrown up from underneath is perfectly refuted. Near the lake Onega it is found reſting on primitive limeſtone, 1 Nev. Nord. Beytr. 137, and ſometimes on argillite, ibid. 149, and the argillite on granite, ibid.; ſometimes on a quartzy gneiſs, 3 Nev. Nord. Beytr. 173.

§ 12.

Serpentine and Pottſtone.

Serpentine is often ſound in layers alternating with granular limeſtone, Berg. Kalend. 215, or underlaying it, 8 Sauſs. 87, and under gneiſs, per Charp. 4 Helvet. Mag. 546; and in his Saxony, p. 398, he ſhews them to be of contemporaneous formation. Zobtenbeg, in Lower Sileſia, conſiſts entirely of ſerpentine, in which ſome hornblende is found; its ſtrata are nearly vertical, 4 Berl. Beobacht. 353. Whole mountains of green ſerpentine are alſo found in Siberia, 1 Berg. Journ. 1791, 99, 102; in ſome it contains white felſpar, and alternates with jaſper, ibid. 103. So alſo mountains of it exiſt near Genoa, where it is called *gabbro*; when it contains veins of white ſpar, it is called *polverezza*, Ferber *Italy*, 118, 119. It forms part of Rabenberg, and is found between gneiſs, Charp. 175. It is not metalliferous, Wern. K. Claſſif. 14; however, magnetic iron ſtone and veins of copper ſometimes traverſe it, Berg. Kalend. 216. The mountain

tain of Skiolſdaſport in Norway, conſiſts entirely of pottſtone, Waller. *de Montibus*, p. 59; and ſeveral occur in Jemptland in Sweden, and ſome in Tyrol, Haiding 63; they repoſe on ſtellſtein or hornſlate. Aſbeſtiform ſteatite is likewiſe found at great heights in the aſcent of Mount Blanc, 7 Sauſſ. 253, and common ſteatites at the ſummit, ibid. 280. Serpentine near the White Sea, has been found ſeated on trap, 1 Nev. Nord. Beytr. 149. and ſometimes on talky quartz, ibid.; ſometimes on a quartzy gneiſs, 3 Neve Nord. Beytr. 173. The mountain of Regelberg in Germany, which conſiſts of ſerpentine, is magnetic, its ſouth ſide attracts the north pole of a magnet, and its north ſide the ſouth pole, 1 Chy. Ann. 1797, 99, 1 N. Berg. Journ. 257.

§ 13.

Porphyry.

Porphyry has been found under gneiſs and alternating with it, and metallic veins traverſe both without interruption or deviation, or any alteration in their metallic contents,

tents, per Charpent. 4 Helvet. Mag. 545; hence he justly considers them as of contemporary formation. It has also been found in the midst of shistose mica, Berg. Kal. 211, and reciprocally gneiss has been found in the midst of porphyry by Widenman, 4 Helv. Mag. 163. Clay porphyry forms whole mountains in Lower Hungary, but hornstone porphyry seldom does, 1 Berg. Journ. 1789, 600; it there reposes on granite, Haiding, 42.

Porphyritic mountains form cones, and porphyry like basalt sometimes covers other hills, per Werner, 1 Berg. Journ. 1789, 607 (this however may be secondary). Porphyry is also often found in a columnar form like basalt. See 2 Berg. Journ. 1790, 325. Haiding, 48.

The *saxum metalliferum* in Hungary, is a disintegrated porphyry, but in the Palatinate it is a disintegrated amygdaloid (mandelstein) that forms the saxum metalliferum, *per Lasius* in 1 Beg. Journ. 1791, 312.

Porphyry generally forms whole mountains, composed of mighty strata; they repose sometimes on granite, 6 Nev. Nord. Beytr. 115; sometimes on gneiss, Haidinger,

dinger, 46. Charpentier mentions a ſort of ſtone which he calls porphyraceous, though he does not aſcribe felſpar to it. It conſiſts, he ſays, of quartz and indurated clay of variable hardneſs, and containing neſts of clay of different colours, or lithomarga. It ſeems rather a ſort of ſand-ſtone, Charp. 69.

In Siberia it never underlays granite, 6 Nev. Nord. Beytr. 115.

The porphyries with a baſe of indurated clay, frequently contain calcedonies, carnelians, amethyſts, or zeolyte, 2 Lenz. 363, Leſke G. 235, 237. That with a *baſis of hornſtone*, often contains galena, black blende, ſulphur pyrites, ſulphurated biſmuth, ſidero calcite, or molybdena, 2 Lenz. 365, Berg. Kal. 211, 2 Berg. Journ. 1793, 214.

That which contains zeolyte has often been miſtaken for trap, Leſke G. 237.

§ 14.

Hornſlate of Charpentier. Shiſtoſe Porphyry of Werner.

This Charpentier reckons among the primitive rocks, 4 Helv. Mag. 547. Renovantz alſo

alſo found it among the primitive rocks of Altai, Renov. 86, and Patrin mixed with granite and hornblende on the borders of the Baikal, 38 Roz. 227, 229. Mountains of it are found in Siberia, Renov. 31; and Schlangerberg chiefly conſiſts of it with interſperſed hornblende and felſpar, ibid. 86.

Alſo in Bohemia, where it ſometimes *repoſes on*, and ſometimes underlays trap, Reuſſ. *Bohemia* XXXII. and III.; the mountain of Schloſsberg conſiſts almoſt entirely of it, 1 Berg. Journ. 1792, 222; that of Bilinirſtein forms pillars like baſalt, ibid. 237, and Reuſſ. M. Geb. 100; and ſome alſo in Saxony, Charp. 29. Sauſſure diſcovered it near Pfaffenſprung intercepted between ſtrata of gneiſs, 7 Sauſſ. 91.

§ 15.

Sandſtone.

This ſtone is generally reckoned among the ſecondary, yet where no organic remains are found in it, where it does not reſt on any ſecondary ſtone, where no ſecondary ſtone enters in its compoſition, I do not ſee why it may not be aggregated

aggregated to the primary. Sand amongſt the convulſions occaſioned by the volcanic irruptions before the creation of animals, muſt have been formed, and even independently of theſe ſome muſt have been depoſited, during or after the cryſtallization of the various ſubſtances contained in the chaotic fluid, ſee 5th Sauſſ. 294. Mount Jorat and the Coteau de Boiſſy, near Geneva, 1 Sauſſ. 246, 349, ſeem to be primeval; ſo alſo the ſandſtone found in the iſland of Bornholm, 5 Berl. Beobacht. Alſo that mentioned in 2 Sauſſ. § 763, which graduates into gneiſs, muſt alſo be primary, though it contains tumblers *(caillous roulés)*. The ſandſtone near Liſchau, in the vicinity of Prague graduates into hornſtone, and even into granite. Mr. Roſler even thinks it to have been originally a granite, whoſe felſpar was decompoſed into clay, which then cemented the quartzy grains; a moſt ingenious and probable conjecture, 1 Bergbau, 339 and 341.

Moſt of the arenilitic mountains of Bohemia, on both ſides of the Elbe, appear to me primitive, by Reus's deſcription, ſee Reuſs

 96,

96, &c. in the E. and N. parts of Bohemia, many of them are ſplit, or form columns reſembling baſalts, 2 Berg. Journ. 1792, 70.

In Bohemia, ſandſtones with an argillaceous cement alternate with thoſe whoſe cement is ſiliceous, Reuſs. In Kinneculla, the loweſt ſtratum incumbent on granite ſeems alſo to be primitive, over it the ſecondary ſtrata repoſe, 29 Swed. Abhand C. 29, 5 Bergm. 126.

In Brainſdorf in Saxony, it paſſes into ſhiſtoſe mica and alternates with argillite, 2 Crell. Beytr. 64. In Riegelſdorf it forms the fundamental rock on which *ſemiprotolite* immediately lies, which is covered by other ſecondary ſtrata, 2 Berg. Journ. 1790, 285. Near Oyben and other tracts of Saxony, no petrefactions or conchylaceous impreſſions are found in it, though in that of Pirna adjoining, they are found, Charp. 24 and 26; it ſometimes repoſes on hornſlate, Charp. 24.

The mountain Steinthal, in the Voſges, of red ſandſtone, is conſidered, by Baron Diedrich, as primeval. 2 Diedr. *Gites des Minerais*, 209, 210. The ſandſtone mentioned

tioned in 6 Sauſſ. 81, which alternates with primitive limeſtone, muſt alſo be primitive.

§ 16.

Rubble Stone.

This alſo is generally thought to be only of ſecondary formation; Voight and Laſius, indeed, judge it to be among the moſt ancient of theſe, yet in the circumſtances mentioned in the laſt ſection, it ſeems to me to rank among the primary.

The rubble ſtone found in Siberia, on the banks of the Iſſet, near Kamyſchenka, mixed with fragments of ſerpentine, ſeems to be primitive; it is not ſtratified, but in huge maſſes, 1 Chy. An. 1793, 512, and 2 Herman, 312.

§ 17.

Farcilite (Pudding ſtone), Breccias and Marlites.

Breccias, conſiſting of fragments of jaſper and flints, cemented by indurated, or rather lapidified clay, occur frequently in Siberia; they are ſtratified, and appear to be primitive, on the banks of the Sentelek, 1 Berg.

Journ. 1791, 83. Whole mountains of marlite are found in the fame vicinity; in fome, breccias, porphyry, jafper, ftratified fandftone, and marlite, alternate with each other; and, therefore, as the Jafper is always primitive, fo muft, in this inftance, all the others, ibid. Alfo an entire mountain of farcilite, containing flints, ftuck in a quartzy bafis, has been difcovered there, ibid. 89.

The argillitic farcilites found by Mr. Sauffure, defcending from the mountain of Balme to the Valorfine, 2 Sauff. § 692. appear to me to be primary; they confift of rounded primeval ftones in an argillitic cement.

A mountain of farcilite exifts in Siberia, near the rivulet *Tulat*, confifting of rounded fragments of jafper, calcedony, aigue marine, and carnelian, in a quartzy cement, 1 Berg. Journ. 1791, 81 and 106. Some farcilites with an argillaceous cement, are alfo found in Dauria, per Patrin, 38 Roz. 22, which he thinks fecondary; but by my definition at leaft, muft be primary, as no petrifactions are found in that country.

At Meifenheim, in the Palatinate, over, and adjoining to vaft maffes of trap, there are immenfe heaps, and even a mountain of farcilite, ſarcilite, 1 Berg. Journ. 1791, 310. The famous mountain of Monſerrat, near Barcelona, ſeems to be a breccia formed of ſandſtone, quartz, and baſanite, in a calcareous cement, by Bowles's deſcription, Bowles, 402. Sauſſure alſo thinks ſarcilites may be primitive, 8 Sauſſ. 290.

§ 18.

Granular Limeſtone.

That this ſtone is primeval, and not formed of ſhells, is now generally agreed upon, ſee Werner, Voight, Charpentier, Sauſſure, Ferber, Herman, &c. no petrefactions are ever found in it, nor does it ever reſt on, or alternate with ſecondary rocks; its height frequently exceeds 7 or 8000 feet, and it often underlays primitive rocks. Whole mountains of primeval limeſtone are found in Stiria, Carinthia, and Carniola, Berg. Kal. 214; and in Swiſſerland, 4 Helv. Mag. 116; and in the Pyrenees, La Peyrouſe, 336; that of Carrara in Italy is well known, Fenſter anhors, Jungfrau, Schreckhorn, in Swiſſerland, all exceed 10,000 feet in height, 4 Helv. Mag. 115, 116; alſo in

Scotland, being probably that which Williams calls mountain limeſtone, 1 Williams, 58; it often preſents lofty ſpires like granite, 8 Sauſſ. 332.

Mountains entirely formed of primitive limeſtone are not commonly ſtratified, but conſiſt of huge blocks without any regular dip or direction, like granitic mountains, Laſius *Hartz*, 60, 175, 176; yet in polygenous primeval mountains it is frequently found in ſtrata; thus in Dauria, near the village of Pechova, it alternates with ſiliceous ſhiſtus, 38 Roz. 298.

At Altenberg, near the lake of Nevenberg it forms ſtrata, 4 Helv. Mag. 306. Mr. Sauſſure on Simplon, found it intercepted between layers of ſhiſtoſe mica, Roz. 7; in Saxony it frequently alternates with ſhiſtoſe mica and argillite, Werner, K. Claſſif. 14, and huge ſtrata of it have been found in gneiſs, and ſhiſtoſe mica and argillite, Charp. 85, 86, 87, 174, 201, 400. Berg. Kal. 214; and between ſienite and hornblende ſlate, Voight, Prack. 48. The unigenous limeſtone mountains of Carniola are ſtratified, but contain no petrifactions, 2 Phyſ. Arbeit. 3, 4, 7. One of the moſt ſingular

ſingular mountains of this ſort is that of Filabres, Bowles's *Spain*, 149. It conſiſts of a block of white marble, three miles in circumference and 2000 feet high, without any mixture of other ſtones or earths, and ſcarce any fiſſures; hence the notion of Buffon, that calcareous ſtones of the ſort we call *primitive*, are of ſecondary formation, and reſult from the tranſudation of lapidific juices from ſuperior ſtrata of conchyferous limeſtone, is evidently groundleſs; where? and how high muſt have been the ſtrata of ſhelly limeſtone from whoſe tranſudation ſuch a mountain as this, or Schreckhorn, could have reſulted? Primitive limeſtone is not always white, nor is the grain of it always very perceptibly ſcaly or lamellar, but approaches by reaſon of its minuteneſs, ſo nearly to the compact, as to paſs for ſuch; nay, it is ſaid ſometimes to diſcover a ſplintry fracture, but very rarely; ſometimes its texture approaches to the fibrous, 1 Nev. Nord. Beytr. 137. It is frequently dark iron grey, or reddiſh brown, both in Siberia and in the Alps, and hence has been taken for trap, as already mentioned. So the mountain of Rauſchenberg on the frontiers of Bavaria is formed of limeſtone ſeemingly compact, filled with ſpar and without petrifactions, 2 Bergm. Journ. 1789, 920. So the weſt ſide of the Ourals is a chain of mountains of compact limeſtone in which no petrifactions are found, or only a few ſcattered impreſſions, per Pallas, 1 Act. Petrop. 48. The mountains of Wetterhorn, Wellhorn, and Burghorn, are primitive limeſtone, whoſe fracture is ſaid to be ſplintry, and their colour greyiſh yellow, 4 Hel. Mag. 272, 273. A ſtratum of it free from petrifactions exiſts in Toſchnitz in the foreſt of Thuringia, but this to me appears ſuſpicious, Voight Prack. 48. As ſome traces of marine acid are found in ſecondary limeſtones, Bergm. Vorleſung. 278, and Georgi in Act. Petropol. 1782, 278, and none are found in primitive limeſtone, this may be a good teſt for diſtinguiſhing them. The ſtratum of yellowiſh and bluiſh grey limeſtone, found in the midſt of a mountain of ſandſtone called Kalkberg in Bohemia, Reuſs *Bohemia*, 117, ſeems to me primitive, though its fracture is ſplintry, paſſing to the uneven.

This ſtone often contains a mixture of various other ſpecies; in Dauria it often contains

contains 50 per cent. of ſilex. 1 Chy. Ann. 1791, 155. being of cotemporary formation. In Switzerland, Schreckhorn, Fenſteraahors, and Jungfrau, conſiſt of limeſtone mixed with flints and minute quartz. 4 Helvet. Mag. 116. In Stiria it is ſo mixed with mica as to reſemble gneiſs, or ſhiſtoſe mica. Werner K. Claſſif. 15. and Wild *montagne ſaliferes* 30; ſometimes $\frac{1}{3}$ of it is argil and quartz, 38 Roz. 232. Sometimes even hornblende, actinolites, aſbeſtus, ſteatites, and ſerpentine, are found mixed with it. Wern. K. Claſſif. 14. Berg. Kalend. 214. Charpent. 400. 1 Nev. Nord. Beytr. 137. and ſhorl in Siberia. 1 Chym. Ann. 1791, 345. Its fiſſures are often filled with ſpar, and ſometimes with quartz. Ferber *Italy*, 450. Renov. xiii. It ſometimes enters in aggregates of the granitic kind, forming ſingular groups; ſee the excellent memoir of Mr. Le Febre on this ſubject in 39 Roz. 354. As 1°. A fragment of quartz, mica and limeſtone, confuſedly cryſtallized. 2°. Again actinolite, ſteatites, quartz, felſpar, and granular limeſtone; another with garnets, both theſe from the Pyrenees; another of ſhiſtoſe mica, with compreſſed ovoidal

ovoidal limeſtone; another from Iceland, in which ſteatites, ſhorl, and cryſtallized mica, are imbedded in the limeſtone, from Sibbo in Iceland; hence there is no ſpecies of ſtone with which our limeſtone may not be aſſociated, or into which it may not enter, or which excludes it: it has been found in the centre of a block of granite, in Tyrole, 39 Roz. 9. and rounded agates have been found in it, in Idria, 3 N. Act. Petropol. 262. Of metallic ſubſtances, few, at leaſt in Europe, are found in it, and theſe are galena, magnetic iron ſtone, and, rarely, black blende. 1 Berg. Jour. 1789, 176. Lately alſo ſome ſcattered ſlips of the black and red ſilver ores, and ſulphur pyrites. Charp. 401. Berg. Kalend. 215. 2 Berg. Jour. 1792, 133 in note. But in Siberia the richeſt copper mines are found in it. 1 Chy. Ann. 1793, 511. It often *repoſes* on *granite*, 1 Chym. Ann. 1793, 510. Renov. 32. Gerh. Geſch. 66. or on *gneiſs*. Charp. 217. or on *argillite*. Ferber *Italy*, 30. 4 N. Act. Petrop. 285. or on *porphyry*. Ibid. On the other hand, it often *underlays*, or is *covered* by, *granite* or *gneiſs*. 1 Chy. Ann. 1791, 155. 4 Helv. Mag. 314 and 546. Charp. 218.

39 Roz. 9. per Dolomieu 1 Berg. Jour. 1789, 174, 175. or *argillite*, Lasius, 178. or primeval trap. 4 N. Act. Petropol. 286.

Near Averdach, Lasius found it between strata of argillite. 1 Berg. Jour. 1791, 306. Saussure found gneiss inclosed between its strata. 8 Sauss. 201. The bluish or reddish grey of Saxony is mixed with argillite. 1 Berg. Jour. 1789, 169; sometimes with hornblende, and then is remarkably fusible. Ibid. 171. In the mountain of *Campo Longo*, not far from St. Gothard, it is found mixed with tremolite, and stratified, 41 Roz. 89. at the height of 6400 feet. When stratified its stratification is always imperfect. 1 Williams, 58. Yet 2 Wms. 22 and 26, it is said to occur frequently regularly stratified.

Marlite.

This also forms whole mountains without any trace of stratification or animal remains, some of which are also metalliferous in Siberia. 2 Herman, 312.

§ 19.

Gypſum, Fluor.

Fluor, in an earthy or ſandy form, has been found at the depth of ſixty feet, in a vein of quartz, at Kobola Poiana in Upper Hungary. Emerling Mineralog. 516. But compact fluor is ſaid to form whole ſtrata in the mountains of the foreſt of Thuringia. Voight Prack. 70. and Werner Enſteh. der Gange 111. and hence he counts it among primitive ſtones.

Gypſum has been found on Mount St. Gothard mixed with mica, 44 Roz. 183. but ſuch inſtances are exceeding rare, and by Pallas it has been found aſſociated with felſpar in Siberia. 5 Nord. Beytr. 280. Baroſelenite is found in conſiderable beds in Savoy. Werner Gange, 111. Might not all theſe combinations take place in the chaotic fluid?

Fluors are alſo found in the cavities of granite at Baveno; in that caſe they are paraſitic. Pini Mem. Surla Cryſtallization des Felſpars, p. 28.

§ 20.

§ 20.

Topaz Rock.

The component parts of this ſingular aggregate I have already mentioned in my firſt volume of Mineralogy; it is reckoned among the primitive, both by Werner and Voight; it forms part of a mountain near Averbach in the metalliſerous mountains (Erzgebirge) of Saxony, but no metal has as yet been found in it. Various other anomalous aggregates daily occur.

§ 21.

Of polygenous primeval Mountains.

Heretofore it was ſuppoſed, that primeval mountains were *entire*; or, if in an undivided mountain, maſſes of primeval ſubſtances occurred, that they formed collateral parts, without either being ſubjected to the other; conſequently, all ſtratified mountains were conſidered as *ſecondary*, and called *undigenous* (flotzgebirge); but of late, in many primeval mountains, ſtrata have been diſcovered of different kinds, ſuperimpoſed one upon the other; nay, in ſome, different aggregates

aggregates of the primeval have been obſerved confuſedly mixed with each other: a few inſtances of both I ſhall here adduce.

1°. In the Solatariſchian mountains in Siberia, argillite, porphyry, and granite, alternate. Renov. 37 *. So alſo in the Grampians in Scotland. 1 Bergb. 399.

2°. The mountain of Gardette, in the ſouth of France, conſiſts of granite at bottom, over that gneiſs, and over that limeſtone, per Schrieber. 36 Roz. 355.

3°. In the circle of Miſnia, granular limeſtone alternates with hornblende ſlate, and is partly mixed therewith, and hence fuſible; in this, ſhiſtoſe, hornblende, felſpar, quartz, and mica, are diſcerned. 1 Berg. Jour. 1789. 171.

4°. At Ehrenberg, near Ilmenau, granite, ſhiſtoſe hornblende, porphyry, and argillite alternate with each other. Voight Prack. 21.

5°. At Schneeberg, argillite alternates with gneiſs and ſlate mica. 3 Helv. Mag. 190, from Charpentier.

6°. Swucku, in the Norwegian chain, conſiſts

* If I underſtand the author right; the text is to me ſomewhat obſcure.

confiſts of petroſilex and porphyry. Waller: *de Montibus*, 59.

7°. Hornſlate often alternates with trap. 3 Helvet. Mag. 241.

8°. In Chalances, in Dauphiné, gneiſs and primitive limeſtone are jumbled, or mixed together. 24 Roz. 381.

9°. Sauſſure deſcribes Mount Jovet, in Savoy, as conſiſting of ſeveral alternating ſtrata of ſtriated ſteatites, ſerpentine, hornblende, and a micaceous and quartzy limeſtone, which is undoubtedly primitive. 2 Sauſſ. 398.

10°. On the banks of the Selinga, in Dauria, the mountains conſiſt of granite, hornblende, and hornſlate, ſometimes alternating, ſometimes mixed, per Patrin, 38 Roz. 227. And near the Ingoda, hornſlate and granite alternate. Ibid. 229.

11°. All the primitive ſtones being in ſome place or other coeval, ſeem ſo mixed as to paſs into each other, though as to their nature totally different from each other; for inſtance, gneiſs and primitive limeſtone. Charp. 201. granite and argillite. Ibid. 283. In the mountain Silberberg in Bavaria, granite, gneiſs, and ſhiſ-

toſe

tofe mica, are confufedly mixed with each other. Flurl. 255, 256.

CHAP. II.

OF SECONDARY OR EPIZOOTIC MOUNTAINS.

§ 1.

Thefe are either marigenous or alluvial.

The diftinctive character of fecondary mountains is the prefence of organic remains either in their natural, or petrified ftate, (or at leaft of their impreffions below the furface) either entering into, and forming the compofition of the ftony fubftances of which thefe mountains confift, or imbedded in them, or lying between their ftrata, or under them; there may be fome, however, as thofe which *evidently derive their origin from the difintegration of primitive mountains*, in which no organic remains may be found, as the carboniferous hills. In *great chains*, the fecondary calcareous are generally on the outfide; within them the argillitic, and the primary in the midft. 1 Sauff. 402. Pallas, Gentil, Ferber, Born, &c. but *elfewhere* they frequently

quently lie promiſcuouſly with each other. The ſecondary, even when otherwiſe independent, always reſt on, and cover, primary; but very commonly alſo they lean on their ſides or inveſt them. 2 Sauſſ. 338. § 919. but they are never covered by primary. 1 Gerh. Geſch. 125. Lehm. flotz. Gebirge, 169. 13 Ann. Chy. 164. 167, &c. Quartz and felſpar in an arenaceous form are often found in them, and alſo mica, as in the ſecondary ſtrata of Mansfeld, 1 Gerh. Geſch. § 72. Between primary and ſecondary ſtrata, commonly a ſtone or ſubſtance of a middle nature, participating of both, intervenes, as Sauſſure has well remarked, as farcilite, or ſandſtone, or breccia. 1 Sauſſ. § 594, p. 528. the reaſon of which I have given in my firſt Eſſay; this obſervation was firſt *generalized* by Mr. Sauſſure: the ſtone itſelf is called *todt liegendes*; I call it, from its conſtitution, *ſemiprotolite*, being ſemiprimigenous. Sometimes petrifactions are found in the upper ſtrata, and none in the lower, 1 Sauſſ. § 409. p. 333. and Laſius. The ſtrata of ſecondary mountains are ſo much the more irregular or inclined as they approach nearer to primary mountains. 1 Sauſſ.

 § 287.

§ 287. Secondary mountains or ſtrata are not all of equal antiquity, thoſe in which ammonites and belemnites only are found, ſeem the moſt ancient; and, next to theſe, thoſe in which madrepores and millepores are found. Some are *derivative*, being formed ſubſequently to the production of organic ſubſtances, and originating from *diſintegration*.

The principal character by which *derivative* mountains are diſtinguiſhed, conſiſts in their exhibiting vegetable ſubſtances, or petrifactions, or at leaſt their impreſſions, or land ſhells, as thoſe of ſnails, or fluviatile ſhells, with either none, or ſcarce any marine remains, though ſome of diluvian origin may exhibit theſe alſo.

Secondary mountains are either formed of one ſpecies of ſtone, or of ſtrata of different ſpecies, one covering or alternating with the other; the former I call *unigenous*, the latter *polygenous*: theſe are commonly *ſtratified*, the former often not. In ſome, different ſpecies are jumbled together, theſe I call *faraginous*; they are by ſome called *tertiary*, as reſulting from the ruins of other mountains, tumultuouſly and promiſcuouſly heaped together. The ſubſtances that form ſecondary

ſecondary mountains are either calcareous, of which there are various ſpecies, or *argillaceous*, or *ſiliceous*, or arenilitic, or *ferruginous*, or *ſaliniferous*, or carboniferous; of both theſe laſt mentioned I ſhall treat of apart. The ſtrata of ſecondary mountains frequently correſpond with each other both in number, ſpecies, and thickneſs, in different mountains, not very diſtant from each other, as in Weſt Gothland, &c.

Although no certain order prevails in the diſpoſition of ſecondary ſtrata, yet there are particular diſpoſitions, which, according to Voight, 2 Berg. 1793, 211, are conſtantly excluded; thus coal is never found under primitive argillite, &c. but he certainly, in his other inſtances, only proves that primitive ſtones are not found over ſecondary. According to Dolomieu, no ſtratum of cryſtalliſed, or granularly foliated ſtone, is ever found in ſecondary mounts. 8 Sauſſ. 284. he muſt except gypſum.

§ 2.

OF CALCAREOUS MOUNTAINS.

Limeſtone.

During the laſt period of the cryſtallization of calcareous ſtones, fiſh were created, hence ſome of theſe mixed with, and diſturbed the cryſtallization; the ſtones, however, which were formed in theſe circumſtances, have a much more compact grain than limeſtones of ſubſequent formation, which were ſolidified by a more partial cryſtallization. In ſome inſtances, vaſt maſſes or banks of coral have their intervals filled up with cryſtallized calcareous matter; the ſtony maſſes thus formed, are, conſequently, not divided in ſtrata, but rather irregularly rifted like primeval maſſes; of this ſort is the mountain of Iberg in the Hartz. Laſius, 180. The petrifactions are ſcarcely diſcernible until the ſtone is ſomewhat withered. Ibid. 214. and ſeveral of the higheſt calcareous mountains of the Vivarois. 1 Soulavic, 13. as that of Vinezac. Ibid. 190. and the rocks of Bidon. Ibid. 192. and ſome in Siberia. 4 Nev. Nord. Beytr. 191.

were

were thus formed. But more frequently this limeſtone forms ſtratified mountains; as at Burren, in the county of Clare; the loweſt ſtrata of theſe are the broadeſt, the ſucceeding ſtrata narrower, ſo that the aſcent is like that of ſtairs; their height is from 600 to 1200 feet above the level of the ſea; no other ſpecies of ſtone interpoſes between them.

In many inſtances, however, thin beds of clay, ochre, and even flints, or hornſtone, (and agates near the Volga. 3 Nev. Nord. Beytr. 162.) occur between limeſtone ſtrata. (2 Herm. 302.) Voight Prack. 109, 110. 11 Ann. Chy. 226. 1 Soulavie, 189. or even in the midſt of a limeſtone rock. 1 Sauſſ. 197. In Bohemia the interceding hornſtone has been found of the thickneſs of five or ſix inches. Reuſs Mittel Gebirge, 38, 39. This argil and the ſiliceous ſtones are evidently the reſult of tranſudation or percolation. In Savoy, Saleve, the Dole, and the Mole, are ſtratified, an inſtructive account of which may be ſeen in Vol. I. of Sauſſure's Voyages. The ſhells found in limeſtone ſtrata are commonly entire; ſome preſerve their internal mother of pearl gloſs,

 and

and do not therefore appear to have been rolled or tranſported, but rather to have been ſucceſſively accumulated in numerous families; and, therefore, not of diluvial origin. In ſome, the calcareous matter is regularly cryſtallized, but more commonly confuſedly; frequently only the impreſſions of ſhells remain; in their rifts calcareous ſpar or granular limeſtone often occurs. Some limeſtones appear to be of late origin, for the ſhells found in them are of common garden ſnails; ſuch occur in Siberia, ſee 7 Kaiſer Accad. Naturforſh, 411. 3 Crell. Chy. Archives, 261. Vaſt caverns are frequently formed in the interior of theſe mountains, or even in plains whoſe ſubſtratum is calcareous rock; theſe evidently reſult from the eroſion or ſolution of the ſofter calcareous particles by water; ſee Pilkington's Derbyſhire, 63. In the Cevennes, conchiferous limeſtone ſtrata at the ſummit repoſe on argillitic, and theſe again on limeſtone whoſe ſhells are of a different kind, 1 Genſanne Hiſt. Languedoc, 260, 261.

Secondary limeſtone mountains always repoſe mediately or immediately on ſome primeval ſtone; thus in Siberia they have

for

for a basis granite, porphyry, or hornblende. Renov. 75, 76. 3 Ann. Chy. 163. In Saxony they repose on granite. Charp. 51. sometimes on granular limestone. 1 Berg. Jour. 1789, 166. sometimes on argillite, sometimes on sandstone. Ibid. In Sweden they often bear on sandstone. 1 Bergm. Erde. Beschrib. 226. in Swisserland they rest on argillite or gneiss. Ferber, Briefe, 9, 10, &c. or calcareous sarcilite, or semi-protolite. 4 Helv. Mag. 116, 117. 1 Berg. Jour. 1790, 176. The calcareous mountains of the Vicentine and Veronese have also a primeval basis. Ferber *Italy*, 44. In Stiria and Carinthia a micaceous argillite. Ibid. 45. or granite, or stelstein. 1 Phy. Arbeit. 106. or in Bohemia on sandstone. Reuss *Bohemia*, xxxi. and also in the Hartz. Lasius, 221. in St. Domingo on granite. 31 Roz. 174. in Austria and Stiria on hornslate, argillite, stelstein, or granite, per Stoulz 1 Phy. Arb. 83. those of Auvergne on granite. 32 Roz. 186. and 39 Roz. 189. in Hainault on argillite. 25 Roz. 85. at Ivry in Burgundy on granite. 2 Mem. Dijon, 1782, 121. 132. On the other hand, this stone *is covered*, sometimes by sandstone as on the N. extremity of Jura. 1 Sauss. 281.

281. 2 Berg. Jour. 1790, 321. and in the Hartz ſometimes by rubble ſtone. 1 Berg. Jour. 1793, 199. or by a calcareous breccia. Ferber *Italy*, 10. In Saxony a bed of it upwards of one hundred feet thick is found covered with clay and ferruginous ſand, to the depth of ſixty or eighty feet. Charp. *Saxony*, 6. Calcareous ſtrata often cover, indiſcriminately, both primeval and ſecondary mountains, as in the foreſt of Thuringia. 2 Berg. Jour. 1790, 315. It often *alternates* with marlite and ſwineſtone. Wern. K. Claſſif. 17.

In the Hartz it is found between rubble ſtone and argillite. 1 Berg. Jour. 1793, 207. In the mountain of Voirons it is incloſed in calcareous grit. 1 Sauſſ. 217 and 248. In Bavaria it alternates with argillite and marlite. Flurl, 44. The ſuperior ſtrata of limeſtone are commonly the *ſofteſt*, the inferior are harder and denſer; the *loweſt* are generally thin. 1 Soulavie. 178. Some contrary inſtances occur in Swiſſerland, but then a different compoſition alſo takes place.

Sometimes calcareous mountains are rent in various directions, and then preſent polygonal

gonal pillars, quadrangular, pentangular, &c. like basalts, as at Ruoms and Bidon, in the Vivarois. 1 Soulavie, 194. 200. In Saxony a rusty limestone is found resembling a sandy indurated clay, this breaks into prisms; it is there called planar. Charp. 49.

In Europe, limestone is seldom metalliferous, the ores most frequently found in it are galena, the sparry iron ore, sulphur pyrites, and copper pyrites. 2 Lenz. 380, and pure sulphur. 1 Herm. 14.

2.

Swine Stone.

This is either compact, slaty, or porous: it commonly contains no petrifactions. Voight Prack, 105. yet that of Kinneculla contains many, per Hermelin Schwed. Abhandl. 1767, 28. but, as he also says, the same stratum contains some common limestone, it is probable it is in that the petrifactions are found; much may also depend on the degree in which the unctuous matter exists in them; Flurl Orogr. Briefe 76. mentions petrifaction to be found in it. 1

Nose

Nofe Forfetz. 75. but *impreffions* of fifh, or of fhells, infects, or leaves, are often found on this ftone. 2 Lenz. 381. it fometimes contains flints. Lafius, 130. and Blumenb. from Voight's Mag. V. B. 1 St. p. 19. mentions a brown fwineftone in which belemnites are contained. Blumenb. Natur. Gefch. 601, and Flurl Bavaria, 78. It commonly repofes on marlite. Lenz. 382. fometimes on compact limeftone. 1 Berg. Jour. 1789, 167. I ftrongly fufpect that fwineftone is in fome cafes a *primeval* limeftone, penetrated with petrol. Mr. Williams thinks petrol is found in all limeftone, in greater or leffer quantities; in Scotland he fometimes found it of the confiftence of an ointment, in the cavities of limeftone. 1 Wms. 235.

3.

Oviform Limeftone.

This is not common; the balls or globules have for the moft part a grain of fand in the middle. 2 Lenz. 381.

4. *Porous*

4.

Porous Limeſtone.

Often contains impreſſions of leaves, and ſometimes wood. 1 Helvet. Mag. 9. In Siberia it ſometimes covers granite, and contains petrifactions. Renov. 27. theſe are evidently of diluvial origin; thoſe found in high ſummits in Peru, are probably contained in this ſpecies of ſtone, and of the ſame origin.

5.

Marlites, and calcareous Sandſtone.

Bituminous marlite commonly bears impreſſions of fiſh, generally of carps; often of marine plants; often alſo native copper, or copper pyrites, or vitreous copper ore; it frequently reſts on ſemiprotolite. 1 Berg. Jour. 1789, 214. Calcareous ſandſtone frequently forms whole mountains, as thoſe of Voirons and Boiſſy in Savoy. 1 Sauſſ. 215. 247. Common marlites often alternate with clay. Voight Fulda, 15. they ſometimes contain gypſum, ſpar, mica, ſand; frequently ſhells, or petrifactions, ammonites,

ammonites, pectinites, tubulites, and impreſſions of leaves of oak. 2 Lenz. 382. flints are alſo found in marl. 1 Gerh. Geſch. 124. In Swiſſerland calcareous ſandſtones, as in Jorat, repoſe on ſemiprimigenous breccia. 4 Helv. Mag. 118. Struvius obſerved, near Cully, a calcareous farcilite. 2 Berg. Jour. 1791, 242. Charpentier obſerved near Prieſnitz, in Saxony, a ſort of marlite that reſembles granite. Charp. 61. it is alſo found of all colours, and beautifully variegated, and regularly ſtratified, in Scotland. 2 Wms. 35.

In Bavaria, marlites, or indurated marl, alternate with ſandſtone. Flurl *Bavaria*, 555. and ſometimes with ſand. 3 Nev. Nord. Beytr. 28. Hills of ſemilapidified marl frequently occur in Carinthia, Carniola, and in the Tridentine, and Venetian territory. 3 Nev. Nord. Beytr. 28. It has alſo been found between ſtrata of limeſtone and of argillite. Ibid.

6.

Chalk.

Chalk ſeems to reſult from the precipitation of calcareous matter, impregnated with as much fixed air as was neceſſary to precipitate it, and no more; the water that held it in ſolution ſeems alſo to have been charged with a ſmall proportion of argil, and a far larger of ſilex in its ultimate ſtate of diviſion; hence the calcareous matter was not cryſtallized, and its interpoſition in a large proportion prevented the ſilex alſo from cryſtallizing. From ſucceſſive precipitations beds of chalk, moſtly in a horizontal poſition, have reſulted; but by the ſucceſſive infiltration of water, the ſiliceous particles were gradually reunited, and formed with the addition of a few of the calcareous and argillaceous genus, thoſe irregularly ſhaped concretions, known by the name of *flints*, which are found diſperſed in *regular* ſtrata through the chalk. 1 Berg. Jour. 1791, 317. 2 Bergb. 384. Their regular ſtratifications proceeds from the uniform denſity of each ſtratum of chalk, through which they were infiltrated: it muſt not be imagined that thoſe found on the ſurface of chalk were always on the ſurface, as the primitive ſurface was long blown or waſhed off; hence petrified ſhells are often found in chalk, and even converted into calcedony.

Laſius,

Lasius, 106. and *vice versa*, from a similar percolation calcareous spar is found crystallized in the same grottos or hollows, in which rock crystal is found, under immense masses of granite. 2 Sauss. 131. Flints themselves often contain petrifactions. Ferber *Italy*, 383. Woodward's Catalogue, and Leske, S. 209, &c.

The petrifactions found in chalk are echinites, pectinites, chamites, mytilites, corallites, &c. 2 Lenz. 396, but metallic substances are never found in it. Werner K. Classif. 19. Berg. Kal. 232. yet in France martial pyrites are said to be found in it, 39 Roz. 358.

7.

Gypsum.

Mountains, but more frequently hills, only of gypsum, occur in different parts of France, Italy, Spain, Poland, &c. many also exist in Asia, in which it is commonly stratified and horizontal, but some are massive. Ferber 4 N. Act. Petropol. 289. Mountains of it occur also in the N. Archipelago, between Asia and America, per Steller, 18 Roz. 41. and to the easy solu-

bility

bility of ſubſtances of this ſort, the diſparition of the intervening continent may in great meaſure be owing. Quartz, calcedony, agates, hyacinths, hornſtone, ſpar, and boracite, have been found in the gypſum; alſo green and grey copper ore, and galena. 2 Lenz. 398. 27 Roz. 70. and ſal gem. and ſulphur in the Ourals. 1 Herm. 14. Petrifactions ſeldom, but impreſſions of leaves and fiſh, have occurred in 1 Berg. Erde. 253. 2 Lenz. 398. Voight Prack. § 59. Townſend found ſome in that of Spain. Vol. I. p. 144. and ſome alſo occurs in the vicinity of the White ſea, in hills of gypſum. 1 Nev. Nord. Beytr. 144. The mountains frequently contain caverns. Laſius, 204. it commonly *alternates* with marl, or marlite, ſometimes with ſandſtone, and, rarely, with limeſtone. Berg. Kalend. 235. Flurl found it between ſtrata of compact limeſtone, without any trace of marl. *Bavaria*, 79. its different families alſo alternate with each other; ſometimes it only forms veins in calcareous mountains. Bowles, 374. ſometimes it is found on the ſummit of a calcareous mountain. Bowles, 131. Where it forms hills or mountains, the

the upper ſtrata are always impure, being fouled by argil, or marl, the higheſt ſtrata being always argillaceous or marly. In Swiſſerland Sauſſure notices a mountain formed of an aggregate of gypſum, ſand, and argil. 2 Sauſſ. 528. In Auſtria rocks of gypſum of a white, grey, greeniſh or red-diſh colour, occur. 1 Phy. Arbeit. 85. It often *repoſes* on ſandſtone. Voight Prack. 107. ſometimes on ſecondary limeſtone. Mem. Par. 1763, 39, in 8vo. ſometimes on marlite. 6 Sauſſ. 37. In the Hartz it is covered by ſecondary limeſtone. Laſius, 233. and ſo almoſt conſtantly in Saxony. Charp. 350. At Aix, near Avignon, it is immediately covered by bituminous marlite, and with piſcine remains, but does not alternate with it. 6 Sauſſ. 41. That it is of modern formation admits of no doubt, after the curious obſervations of Werner in his letter to Leſke. 2 Biblioth. du Nord. 73 and 82. yet Sauſſure found it in a ſhiſtoſe form mixed with mica near St. Gothard. 7 Sauſſ. 178.

§ 3.

Argillaceous.

Under this head I comprehend not only ſtones, &c. of the argillaceous genus, but alſo ſuch ſandſtones, porphyries, breccias, as bear an argillaceous cement.

I.

Argillite.

There can be no doubt but argillite is frequently of ſecondary origin; Ferber acknowledges it to be partly primeval, and partly ſecondary. 4 N. Act. Petropol. 289. Gruner found ammonites in the argillite near Meyringen in Swiſſerland. 3 Helv. Mag. 191. in a ſpecimen from Heſſia, mytilites occur; ſee Leſke, G. 339. Voight found a limeſtone with petrifactions, between ſtrata of argillite. 1 Mineral. Abhandl. 86, 87, 88. It often contains piſcine remains betwixt its laminæ, Laſius, 105. Sauſſure found argillitic ſtrata intermixed with black marble. 1 Sauſſ. 401. In the Hartz impreſſions of reeds, ruſhes, and pectinites, are found on it where it adjoins to rubble ſtone. Laſius, 103. 105. Sometimes it hardens and grows more ſiliceous from the bottom upwards. Laſius, 105. Sometimes it is harder at greater than leſſer depths. Idem, 102. In the Hartz it *alternates* with, and ſometimes is intimately mixed with, rubble ſtone. Laſius, 138. It alſo paſſes into ſandſtone. Idem, 105. At Kinneculla it alternates with aluminous ſlate and marlite. 29 Schwed. Abhandl. 26.

2.

Indurated Clay.

Pectinites, chamites, mytilites, turbinites, &c. have been found in it, alſo ſpar, gypſum, quartz, ſulphur pyrites, martial ochre, common ſalt, vitriol, and alum, 2 Lenz. 403. It often alternates with limeſtone, gypſeous, and ſandſtone ſtrata. Lenz. Ibid. In Saxony it has been found alternating with thin beds of argillaceous iron ore, filled with ſhells. Charp. 8. Is not the ſtriped jaſper 1 Berg. Jour. 1791, p. 97. in Siberia, in which petrifactions are found, rather an indurated or lapidified clay?

3. *Shale*

3.

Shale (Shiefer thon).

Impreſſions of reeds or fern, ſometimes of muſculites, mytilites, &c. are found on it. 2 Lenz. 378. Alternates with coal and ſandſtone. Voight Prack. Paſſes into argillite, marl, ſandſtone, and rubble ſtone. 2 Lenz. Ibid.

4.

Bituminous Shale (Brandſchiefer).

This is generally found in ſtratified mountains that contain argillaceous iron ſtone. It forms thick ſtrata which alternate with that ore, and with ſhale, marlite, or calamine, and many petrifactions. 1 Emmerl. 291. and Werner's Letter to Leſke in 2 Biblioth. du Nord, 81.

5.

Indurated Lithomarga.

It ſometimes bears the impreſſions of reeds; and is found ſtratified in coal mines. Emmerl. 357. 359.

It paſſes into indurated clay, or mountain ſope, or ſteatites. Emmerl. Ibid.

6.

Rubble Stone, or Slate.

Petrifactions, as hysteriolites, mytilites, &c. are found in it, or impressions of reeds, or rushes, and madrepores, at the depth of 1200 feet. Lasius, 206. Leske, G. 874. Wern. K. Classif. 18. It sometimes covers conchiferous limestone. 1 Berg. Jour. 1793, 199. It is frequently metalliferous; the substances of that sort it bears are galena, copper, and sulphur pyrites, bismuth and cobalt. 2 Lenz. 391. It alternates with, and frequently passes into, argillite and sandstone. 2 Lenz. Ibid. A siliceous sandstone mixed with argillite at Bornholm, is called a rubble stone, 5 Berl. Beob. 94. hitherto it occurs principally in the Hartz.

7.

Argillaceous, or ferruginous Sandstone.

Sandstone, with an argillaceous or ferruginous cement, frequently forms whole stratified mountains in Saxony and Bohemia. Charpent. 41, &c. the strata are often of considerable thickness, and horizontal; the uppermoſt and lowermoſt often the looſeſt; the interior of each ſtratum the hardeſt. Ibid. 44. The loweſt parts contain petrifactions. Ibid. The cement is argillo-ferruginous. In Normandy the ferruginous ſandſtone, called *rouſſier*, conſiſts of calcareous gravel in a ferruginous cement; it contains oſtracites, Mem. Par. 1763, 404, in 8vo. 81, in 4to. but more commonly the granular part is of the ſiliceous genus. Werner, 17.

The petrifactions found in ſandſtone are moſt commonly orthoceratites, chamites, tellinites, oſtracites, &c. and often the impreſſions of leaves and fiſh. Charp. 44, 45. 47. 2 Lenz. 389. hence it cannot be a diſintegrated granite, as ſome have imagined. It ſometimes contains ſhells converted into calcedony. Laſius, 225. 3 de Luc, 15. It is ſeldom metalliferous, Werner, 18. yet the vitreous copper ore, malachite, cinnabar, ſulphur pyrites, iron ſtone, ſometimes occur in it. 2 Lenz. 389. Laſius, 146. But query, if in the ſecondary? In Saxony its loweſt ſtrata *repoſe* on granite, or argillite. Charp. 45. in Bohemia chiefly on granite. Reuſs Bohemia, 100. 103. 110, &c. yet I ſtrongly incline to think the Bohemian ſandſtone, mentioned by Reuſs, primeval. In Swiſſerland it ſometimes leans or reſts on ſecondary limeſtone, as at Mount Ementhel, 4 Helv. Mag. 118. or repoſes on it. 2 Mem. Lauſ. Part II. 17. 5 Sauſſ. 283. In the territory of Ilmenau, Voight found it repoſing on ſecondary limeſtone. Voight, 53. 55. It *paſſes* into argillite, rubble ſtone, or wacken. 2 Lenz. 389. Laſius, 145, 146. 151. It alternates with unguilite (Nagel fluhe) in Swiſſerland 1 Berg. Jour. 1790, 174. and in Bavaria. Flurl, 88. In the Hartz it forms the higheſt ſtratum of moſt of the mountains, Laſius, 148. covering argillite. Id. 150. yet this ſtratum contains no petrifactions, though the lower ſtrata do; but this is mere chance. Mem. Par. 1747, 1059, in 8vo. Ibid. Sometimes it covers ſwineſtone. Id. 225.

When ſomewhat withered, the ferruginous part often fails, then it appears porous; yet theſe pores often contain iron in an ochry form, which ſufficiently indicates their origin. Laſius, 149.

8. *Porphyry.*

8.

Porphyry.

In Bohemia ſhiſtoſe porphyry ſeems to be ſometimes a ſecondary rock, for impreſſions of leaves have been detected betwixt its laminæ. Reuſs Mittel Gebirg. 101, 102. In the Hartz porphyry alternates with a red argillite, which is ſecondary; for a petrified nautilite was found in it. Laſius, 154. 110. In Saxony it ſometimes covers a ſecondary limeſtone, and therefore muſt be ſecondary. Charp. 49. who remarks that it bears a ſtrong reſemblance to granite. Ibid. 50.

9.

Trap, and Baſalt.

The conſtitution of this ſtone has been already ſufficiently diſcuſſed in the firſt chapter: we have there ſhewn, alſo, that in ſome places it is of primeval origin; but now it will be made to appear that it is often, alſo, of ſecondary origin, as it will be ſhewn to repoſe on ſtones evidently of that origin.

The mountain of Kinneculla, in West Gothland, is stratified: its highest stratum is trap; the next to that is argillite; and to this reddish brown compact limestone: under these swine stone, and a mixture of gypsum and swine stone, and a grey slaty limestone, in which petrifactions are found: below this sandstone; and under that granite. Swed. Abhand. 1767, 25, &c. Petrifactions are also found in the compact limestone. Id. and 5 Bergm. 124. Hence, as this trap is incumbent on secondary stones, it must also be itself secondary; and Faujas, approving of Bergman's reasons, allows this trap not to be of volcanic origin; see his tract on traps, p. 12. and yet it is *sonorous* like basalts, which on that account has been thought volcanic. Some wedish mineralogists imagined the trap was only *invested*, but not *underlayed*, by the other strata, apparently inferior to it, but both Bergman and Hermelin have proved the contrary.—To demonstrate the falsehood of this opinion more fully, we have only to observe, that at Pohlberg, in Saxony, where basaltic pillars repose on gneiss, a gallery has been pushed into the gneiss, under the basalt,

basalt, without meeting it, so that it is plain it had not pierced up through the gneiss, but was barely laid over it, per Charpent. 223. 4 Helv. Mag. 546. 1 Berg. Jour. 1789, 258. Werner observed the same at Anaberg in Saxony, where the basalt reposed on clay, into which a gallery was worked without meeting the basalt, though under it. 2 Berg. Jour. 1788, 851. and Faujas in the . . . . , near Mezinc, where the pillars repose on coal. Recherches sur les Volc. Eteints, 338.

In Hessia the basaltic or trap mountain, called *Noll*, reposes on blue compact limestone, which contains petrifactions, as Mr. Riess, who lately described it, observes. 1 Berg. Jour. 1792, 373. But the most considerable basaltic mountain in Hessia, is that called Meissener; it is above 1200 feet high, the summit is formed of basalt, but the inferior strata are coal, and bituminous and carbonated wood. 1 Berg. Jour. 1789, 274. 1 Berg. Jour. 1792, 378. This trap or basalt must then be secondary.

In Bohemia the basalt of Luschitz rests partly on compact limestone, partly on argillaceous marl. Reuss M. Gebirg. 72, 73.

In

In Scheibenberg the basalt passes into wacken, which passes into clay; the clay into quartzy sand, and this lies on gneiss. 2 Berg. Jour. 1788.

In the environs of Sortino, in Sicily, Dolomieu found trap covering a limestone that abounded in shells, a circumstance that embarrassed him exceedingly, as he thought it a lava, but could not discover from whence it issued. 25 Roz. 192. Basaltic pillars are sometimes of an extraordinary length, those of Stolpe, in Saxony, exceed 300 feet without any articulation or division, Charp. 37. and in the Isle of Feroe, 120 feet long, and 6 feet in diameter. Nose Samlung. 138. Basaltic pillars are sometimes dilated in the middle, and incurvated; this dilatation and curvature Soulavie discovered to proceed from a *nucleus* of a foreign stone in the part dilated; yet frequently this nucleus was found only in one pillar; but the bending of that pillar obliged, by its pressure, the next pillar to bend also, as did this the next, and thus the whole range of them was incurvated. This observation evidently proves that all the pillars were originally in a soft state, and formed a coherent mass, for if they were originally as distant from each other as at present (in the mountain of Antraigues, where he made this observation), they could not press one on the other, and, consequently, that these pillars did not assume the columnar form by crystallization, but by disruption, as I have elsewhere shewn. It also deserves to be remarked, that the granitic *nucleus* did not appear in the least altered; a sign, surely, that it had experienced no heat, and, consequently, that the softness of the basalt did not proceed from its having been in a state of fusion. 2 Soulavie, 76 and 51.

In a mountain of Bohemia, basalts have been found standing on clay, which formed $\frac{1}{3}$ of the whole mountain; and, therefore, could not have arisen from the decomposition of the basalts, as some have imagined clay to do in a similar case. Reuss *Bohemia*, 177. Impressions of leaves or shells have sometimes been found on it. Lenz. 374. or on the marl contained in its cavities. 1 Chy. Ann. 1792, 70. and Bruckenman found muscle shells, and ammonites, and corallites, in basalt of the pretended extinct volcanoes of France. 1 Chy. Ann. 1794, 103.

103*. It is found imperfectly stratified in various parts of Scotland. 1 Williams, 58 and 71. He observed columnar basalts to be perfectly vertical, when the stratum they reposed on was horizontal; but oblique if the stratum was inclined. 2 Williams, 38. On the western declivity of Snowden, basaltic pillars rest on hornstone. Aikin *Wales*, 99.

10.

Hornblende.

This has very unexpectedly been proved to be sometimes of secondary formation, for Baron Ash has presented to the museum of Gottingen a number of shells, mostly tellinites, filled with striated shining hornblende, found in the Crimea. Blumenb. Handb. Natur. Gesch. 579, in note.

11.

Argillaceous and calcareous Breccias, and Farcilites.

Lasius observed, argillite, quartz, and hornstone. stuck in an argillaceous cement, forming a rubble stone breccia in the Hartz; its specific gravity 2,579. Lasius, 143, 144.

* Doctor Richardson lately discovered, and shewed me, shells in the basalts of Ballycastle.

The

The hill of Morsfeld, in the Palatinate, in which mercurial ores are found, is composed of vast layers of argillaceous breccia, betwixt which thin strata of argillite are intercepted, per Lasius, 1 Berg. Jour. 1791, 308; on these impressions of fish are discerned. Ibid. 309. Whole mountains of calcareous breccias are met with at the foot of the Suabian Alps, in Swisserland (the *Nagel fluhe*), and Italy (*Marmo Brecciato*). 2 Widenman, 1036. A calcareous farcilite, there called *amenla*, is found in France, extending six miles from Montmoirac to Rousson; it consists of heaps of calcareous rounded pebbles agglutinated by a calcareous cement, and intermixed with shells; it does not form strata, but rather immense heaps; the upper pebbles are not so perfectly cemented as the lower. Mem. Par. 1746, p. 723, &c. Another of the same kind in Carniola. 2 Phy. Arbeit. 8. (See p. 97.)

Also the mountain of Rigeberg is formed of tumblers of limestone and hornstone and a red argillaceous stone cemented by a calcareous cement. 7 Sauss. 188, and Monserrat. (See Bowles.)

§ 4.

§ 4.

Siliceous.

1. *Hornſtone.*

According to Baron Born, petrifactions have been found in a white hornſtone near Lehotka, in Hungary. Letter 21, p. 212. In the peak of Derbyſhire it is ſaid to be found in ſtrata twelve feet thick, even not always ſtratified,—and alſo in limeſtone. 6 Phil. Tranſ. Abr. Part II, 192. Petrifactions are found in it, Whitehurſt, 233, 234. Hills, conſiſting of hornſtone, penetrated with calcareous particles, and paſſing into marlite, have been obſerved in Auſtria. 1 Phy. Arbeit. 80, 81. Another quarry of this ſtone, with petrifactions, exiſts in the Dutchy of Luxemburg, deſcribed by Noſe, 1 Chy. Ann. 1789, 425. That alſo which is ſo frequently intercepted betwixt beds of compact limeſtone, muſt be ſecondary; ſee 6 Sauſſ. 82. It is itſelf often the ſubſtance of petrifactions. Flurl *Bavaria*, 571.

2.

Jaſper.

Madrepores are ſaid to have been obſerved in red jaſper, in Hungary. 2 Lempe. Mag. 76. Alſo in the red jaſper of Breſcia, Ferber *Italy* 33. This Bergman alſo mentions in his Phyſ. Geo. 304, but juſtly queſtons it. Laſius, p. 207, tells us he found a nautilus petrified in red jaſper, near Elbingerode. Striped jaſper with petrifactions, is ſaid alſo to be found in Siberia. 1 Bergm. Jour, 1791, 97. Herman doubts it: I ſuſpect that in all the ſuppoſed inſtances of ſecondary jaſper, the word jaſper has not been taken in the ſtricteſt ſenſe.

3.

Farcilites, and Breccias.

A ſort of breccia formed of iron nails agglutinated by a ſiliceous cement has been taken out of the Danube, and another out of a canal in Hungary. 1 Phyſ. Arbeit. 79. In Siberia whole mountains of ſiliceous farcilites are found conſiſting of rounded pebbles of jaſper, calcedony, and carnelian,

carnelian, ſtuck in a ſiliceous cement. 1 Bergm. Jour. 1791, 81. Some alſo of ſiliceous breccias. Ibid, and in the ſequel. I mention theſe here, as it may be doubted whether they are primeval or not. Farcilitic mountains are alſo common in the north of Scotland, and the Weſtern Highlands. 2 Williams, 51.

4.

Siliceous Sandſtone.

In the ports of Domich and Campara, in the iſle of Arbe, on the coaſt of Dalmatia, Abbé Fortis found oſtracites in a ſiliceous ſandſtone. *Dalmatia* 355. See alſo in Leſke, G. 793, &c, ſeveral petrifactions in this ſort of ſandſtone. The hill of Platenberg conſiſts of ſandſtone with a calcedonic cement; it contains ſhells converted, by expulſion of the calcareous matter, into calcedony. Laſius, 284, 285.

5.

Semiprotolites (Rothe tod liegendes).

Theſe ſtones I call by this name, as being partly of primeval, and partly of ſubſequent, origin; they conſiſt of pebbles, or of fragments,

ments, or of ſand of primeval origin, compacted and cemented by an argillaceous or calcareous, or ſiliceous cement of poſterior origin; hence they generally form the loweſt ſtratum that ſeparates primeval rocks and ſecondary ſtrata. From their compoſition they come under the denomination either of *farcilites, breccias,* or *ſandſtones.* In ſome places this ſand has been accumulated into vaſt heaps ſo as to form mountains 6 or 700 feet high, and then compacted by an adventitious cement. Of this ſort are the mountains of Hertzberg, and Kaulberg near Ilefeld, in which the ſand is cemented by a ferruginous cement, and contains fragments of porphyry, and alſo veins of iron ſtone, and manganeſe, and ſtrata of coal, with impreſſions of reeds, ruſhes, and other plants, Laſius 249, and 280. The red colour is evidently from iron.

The ſemiprotolite of Wartburg near Eiſenach, contains rounded lumps of granite and ſhiſtoſe mica: ſubſtances found in the neighbouring mountains. The ſemiprotolite of Goldlauter conſiſts entirely of porphyry, as do the primeval mountains of that diſtrict. That of Kiffhauſerberg in

S Thuringia,

Thuringia, contains rounded argillites from the neighbouring mountains of the Hartz. Petrified wood is found in this laſt. Voight's Letters, 19, 20. According to Voight, the ſemiprotolite found under coal has a ſiliceous cement, and contains few primitive ſtones, *Lettres ſur les Montagnes*, 31. Sauſſure made the ſame obſervation on thoſe which he found on the deſcent of Trient, which interceded between the primary and ſecondary mountains. 2 Sauſſ. § 699. He even remarked long before, that primeval and ſecondary rocks were almoſt always ſeparated by a ſandſtone or farcilite, 1 Sauſſ. § 594. Where the ſecondary ſtrata are calcareous, the ſemiprotolite has a calcareous cement, See Lehm. 168. Semiprotolite is always red, by reaſon of the ferruginous particles by which it is cemented; its diffuſion or expanſion is unequal, being frequently horizontal or even, but ſometimes depreſſed, and in other inſtances much elevated. Moſt of the ſuperimpoſed ſtrata partake of this inequality, and are its natural conſequences. Hence the protuberances and depreſſions, otherwiſe called *moulds*, obſerved in them. Charp.

Saxony,

Saxony, 371. It reſts on granite. Ibid, 370, 371.

§ 5.

To form a clear idea of the ſtructure of polygenous ſecondary mountains, I ſhall exhibit an enumeration of the ſtrata diſcovered in ſome of them in various parts of Europe, with ſome, though not an accurate deſignation of their thickneſs, as abſolute preciſion is not here neceſſary.

Of the ſtrata from Ilefeld to Mansfield, and round the Hartz. Lehman, 163. Laſius, 278, and Gerhard, 109.

Enumeration of Strata.

		Fathoms.	Inches.
1	Vegetable earth		more or leſs
2	Swine ſtone	6	
3	Gypſum .... from 4 to	30	
4	An arenaceous compound of. clay, chalk, and ſand, from 12 to	20	
5	Aſh grey compact limeſtone ...	2	
6	Argilliferous limeſtone	½	
7	Induraed clay		1
8	Calciferous clay	¾	
9	Calciferous clay ſlate		16
10	Black ſlightly cupriferous marlite		6
11	A blacker ditto, and poor in copper		1

		Fathoms	Inches
12	N. 10 repeated		4
13	Rich bituminous marlite		1
14	Bituminiferous and cupriferous sand		1
15	Calciferous and argilliferous gravel	½	
16	Blue clay .... from		2 to 8
17	Red friable ferruginous sandstone mixed with clay and mica ...	1	
18	Red hard semiprotolite from ... 20 to	60	
19	Red siliceous and ferruginous .. sandstone	16	
20	Sandy kraggstone............	¾	
21	Ditto	1	
22	Brownish red slaty wacken	8	
23	Liver-coloured ditto	8	
24	Bluish black ditto	10	
25	Clay slate..................	¼	
26	Coal	¼	
27	Clay slate, with impressions of leaves, and reeds	½	
28	Black slaty trap	15	
29	Red semiprotolite	30	
30	The primeval rock		

Note 1. According to Lehman, the strata after No. 18 are exceptions to a general rule, namely, that semiprotolite should be the last stratum; but he says, that in this case he found strata under it; I believe,

I believe, however, that it is not an exception, but that the succeeding strata were indeed lower, but not exactly under the 18th, for it is not by boring, but by levelling the external extremities of the strata that he judged of their superimposition. Lehm. 169.

Note 2. The denominations were frequently corrected from Lasius, and the specimens Lasius referred to, and sometimes from Gerhard.

Strata of Derbyshire.

Whitehurst's Theory, 182.

		Yards
1	Coarse sandstone	120
2	Slate clay	120
3	Conchiferous lamellar limestone, the laminæ intercepting thin strata of shale	50
4	Amygdaloid (toadstone)	16
5	Compact limestone, like No. 3	50
6	Amygdaloid	46
7	Lamellar limestone, like No. 3	60
8	Amygdaloid	22
9	Limestone as the former	Not yet cut through

Note 1. In the limestone, No. 3, ores of zinc, pyrites, compact baroselenite, fluors, and petrifactions are found.

2. The ſtrata of amygdaloid are progreſſively more ſolid as they lie deeper.

3. The ſtrata of limeſtone and amygdaloid are ſeparated by ſtrata of light blue clay with a tinge of green (probably wacken) of from 1 to 6 feet thick. Theſe contain nodules of ſpar and pyrites, and it is remarked that from them the hot ſprings flow. They are alſo ſtrongly calciferous. Ibid. 187.

Order of the Strata near Bouillet, from Struvius,

2 Mem. Lauſ. 52, and 2 Bergm. Jour. 1791, 246.

1 Mould
2 Puddingſtone
3 Indurated marl
4 Sandſtone
5 Indurated marl
6 Limeſtone
7 Clay
8 Gypſum
9 Grey rock (a mixture of gypſum, clay, and ſand.) p. 28
10 A calciferous, and ſomewhat indurated clay, which he calls *Pierre de corne*
11 Granite

Of Kinneculla.

29 Schwed. Abhand. 27 per Hermelin.

0 Mould
1 Trap
2 Argillite, with thin layers of marlite and aluminous ſlate
3 ſtratified compact limeſtone
4 Swineſtone and aluminous ſlate
5 Sandſtone, alſo ſtratified
6 Granite

CHAP.

CHAP. III.

OF VOLCANIC MOUNTAINS.

Volcanic Mountains are generally of a conic form, yet all conic mountains muſt not therefore be deemed volcanic; for even granitic mountains (as *Achterman* in the Hartz. Laſius 17,) to ſay nothing of porphyritic mountains, often preſent that ſhape.

The internal primordial ſtructure of volcanic mountains is much more difficultly diſcoverable than that of any other ſpecies of mountains. Theſe laſt, with few exceptions, have remained materially unaltered for ſeveral centuries; the former, at leaſt ſuch of them as are beſt known to us, have been convulſed, and eviſcerated ſeveral times, even in the ſame century. Thus their ancient component ſtony maſſes lie buried under their modern adventitious ejections. Neptunian mountains have been perforated to great depths for the purpoſe of extracting metallic ores, and their contents ſtrictly examined by perſons accurately ſkilled in mineralogy; but as lavas contain no metallic veins, and as even ſome danger has been apprehended in excavating them to great depths*, they have only been ſuperficially obſerved, and often by perſons not only deſtitute of competent mineralogical knowledge, but who have openly profeſſed either their ignorance or their contempt of it. One of them expreſsly tells us, that, " Nomenclators have " for a long time *retarded* the progreſs of " natural geography."—" The mind of a " nomenclator is conſtantly warped by his " ſpecimens, he obliges nature to conform " to the ſyſtem of his cabinet. But the *man* " *of genius*, to whom nomenclature is no- " thing," (that is, who does not regard the eſſential differences of various ſubſtances) " compares ſubſtances in their reſpective " ſituations, ſtudies their alteration, com- " bines theſe facts with each other, and " obliges nature to diſcover what ſhe was, " by what ſhe at preſent is, &c.†" The *man of genius* whom he here quotes was ſo ill inſtructed, as to confound pottſtone and ſhorl, and granite and lava. The con-

* Collini Lettres ſur les Montagnes, p. 45.

† 4 Soulavi, 306.

temner of nomenclature himſelf tells us that *granite often conſiſts of quartz ſingly*, and that *quartz does not reſiſt the action of acids* *. The opinion at preſent moſt generally received concerning the ſtructure of volcanic mountains, is, that they conſiſt ſolely and entirely of volcanic ejections: to me, however, it ſeems highly improbable. The volcanoes with which we are beſt acquainted are Veſuvius and Ætna; of Hecla we know but little, and ſtill leſs of the Aſiatic and American volcanoes; now with reſpect to Veſuvius we may obſerve,

1°. That it is a mountain of thirty miles in circumference, and above three thouſand ſix hundred feet in height. In the year 79 of our Æra it was at leaſt as high as that at preſent called *Somma*, which nearly ſurrounds it: ſince the year 79 of our Æra, in which its moſt conſiderable eruption happened, the firſt recorded in hiſtory, there were about 29; of theſe, that of 1631 and 1794, were the greateſt: that of 79 conſiſted chiefly of ſand and aſhes, and produced over the town of Pompei an elevation of about eighty feet; in the laſt

* 1 Soulavie, 454, 455.

eruption,

eruption, the lava was a mile in breadth, and at moſt twelve feet deep, ſo that the mean height produced by each of theſe thirty eruptions cannot certainly be deemed to exceed twenty feet; then in 1700 years theſe eruptions produced an elevation of the mountain amounting only to ſix hundred feet. In antecedent times, or at leaſt for 1000 years before, the eruptions, if any, muſt have been inconſiderable, ſince they are mentioned by no hiſtorian or poet. What reaſon then have we to think that the enormous bulk and height of the whole mountain conſiſt of an accumulation of eruptions ſtill more ancient, and of which we have no account? The lava found under the ſea furniſhes certainly none, as lava has frequently been known to flow into the ſea.

2°. Many reaſons may lead us to think that the internal *nucleus* of Veſuvius is calcareous; though probably under the calcareous, vaſt maſſes of trap, hornblende, and argillite, alſo exiſt, with abundance of ſulphur pyrites, coal, and bitumen. Ferber remarks that the volcanic country round Veſuvius is entirely calcareous. Gioeni aſſures us that $\frac{3}{4}$ of the ejections are of that nature,

nature, p. xxxiv. He adds, that primeval ſtones are found, not indeed near the crater, nor on the exterior beds of lava, but on the ſtratum of mould which clothed the mountain before the cone was formed, and under *all that appears volcanic*; and though, through prejudice, he ſuppoſes them ejected by the volcano, yet he ſays the date of ſuch ejections precedes that of all the lavas that can at preſent be diſtinguiſhed, p. xxxvi, and vii; nay, more, lamellar ſwineſtone has been found in horizontal ſtrata, p. 13. Padre Torre is of the ſame opinion, but his ſentiments, through his ignorance of mineralogy, has been treated by the more zealous volcaniſts either with pity, or with contempt; yet Ferber, one of the moſt eminent mineralogiſts of this century, alſo thinks that limeſtone exiſts under the volcanic matter, adding, that large maſſes of it with calcareous ſpar have been ejected, Letter XI, p. 165. Near Salerno he diſcovered argillite under the limeſtone, p. 166. See alſo Strange, in Phil. Tranſ. 1775, p. 28.

Of the ſtructure of Ætna I have already treated in Eſſay III. I ſhall only add, that its

its height exceeding 10000 feet, and its circumference 130 miles, that circumſtance alone renders it incredible that its whole bulk is volcanic. Though it flamed ſome centuries before our Æra, yet there is no reaſon to think it did ſo before the age of Homer: and how little it has been altered by its numerous ſubſequent eruptions, appears from the ruins of an ancient temple, which, though the temple was built long before our Æra, are ſtill extant and uncovered. As its lavas are moſtly porphyritic, there is great reaſon to ſuppoſe that its primitive baſis was a porphyry, with vaſt beds of coal, pyrites, and bitumen, interpoſed.

The reaſon that ſeems principally to have induced philoſophers to believe that volcanic mountains owed their origin to ſubterraneous fires, was the production of *monte nuovo*, a mountain of conſiderable height, thrown up by a volcano in forty-eight hours; but it ſhould be conſidered, that this mountain differs in its ſtructure alſo from all other volcanoes, for it conſiſts, not of beds of real, or ſuppoſed, lava, but of indurated volcanic aſhes, pumice, and fragments of lava, intermixed with each other,

Diedrich's

Diedrich's note on Ferber Italy, p. 169. a ſtructure very different from that of Veſuvius, Ætna, or Hecla.

The internal heat of volcanoes ſeems much lower than is commonly believed, if we may be allowed to eſtimate it either from the circumſtances of its focus, the weakneſs of its effects, or the low heat of the ejected lava: this heat ſeems to have originated from the decompoſition of water by pyritous maſſes, particularly by that ſpecies of pyrites that contains iron nearly in its metallic ſtate; the ſulphureous part of the pyrites being excluded from the contact of pure air, muſt have been fuſed by this heat, and its fuſion promotes the ignition and fuſion of iron in a very moderate heat, as Van Diemau and his aſſociates have proved; thus the beds of coal were heated, and the bituminous ſulphurated maſſes of ſtone and earth gradually ſoftened and liquified by the fuſed ſulphur and bitumen contained in them, but all this while no flame can be ſuppoſed to exiſt, for want of pure air; or only momentary flaſhes unattended with any increaſe of heat, as in Diemau's experiments: at laſt, however, the ſemiliquified maſs, mafs, after violent ftruggles and efforts, is propelled by the claftic vapours to the crater of the volcano, where commonly meeting lefs refiftance, it rifes, there meeting atmofpheric air, the fulphureous and bituminous vapours are at laft fired with dreadful explofions, and immenfe volumes of flame; the coal is incinerated, and the liquified matter or lava finally expelled. It is only then at the mouth of the volcano that the greateft heat is produced; but great as it may be, the mafs of inflamed matter being incomparably fmaller than that of the lava, the increafe of heat refulting to the whole of this muft be very moderate. Hence we fee, why eruptions that happen after long intervals of reft are by far the moft violent, and attended with the greateft heat, as the air in the crater, and, perhaps, alfo in the interior cavities, becomes, during fuch intervals, much purer, and, confequently, more capable of exciting heat.

In treating of the origin of bafalt, I have given many inftances of the low heat experienced by lavas. Sir William Hamilton, in his account of the late dreadful eruption of Vesuvius, on the 15th of June, 1794, tells us, that the town of Torre del Greco was so quickly surrounded with red hot lava, that the inhabitants thought they had scarcely time to save their lives, yet several of them, whose houses were thus surrounded while they were in them, saved themselves by coming out of their tops the next day. It is evident, that if this lava had been hot enough to melt even the most fusible stones, these persons must have been suffocated; nay, though the lava continued red hot the next, and even on the succeeding days, yet on the 16th many walked over it; see Phil. Transf. 1795, p. 86.

With respect to ancient volcanoes now extinct, there is scarce any part of Europe, or even of the globe, in which they have not been thought to exist, for trap and basalt have been found almost in every country; and since the year 1767, these have been by many supposed indubitable testimonies of an ancient volcano. In France, Languedoc, Provence, but particularly, Vevarois, and Auvergne, have been supposed torn up by numerous volcanoes; many, however, have been by subsequent less prejudiced

judiced obſervers effaced off their liſt *. One of the moſt remarkable, the *Coupe de Aiſa*, is very accurately deſcribed by Mr. Faujas: he takes the black trappoſe matter that deſcends from the ſummit of the mountain to a torrent at its foot where it forms pillars, to have been a ſtream of lava. To a German mineralogiſt who lately examined that mountain, it appeared in a very different light; he tells us, that the whole mountain is formed of trap, and covered by vegetable earth; that a current of water laid bare the part which to Mr. Faujas appeared a bed or ſtream of lava, but did not eſſentially differ from any other part of the mountain; ſee Haiding. Gebirg's Arten, p. 38, in the note.

As to the trappoſe maſs, called *roche rouge*, mentioned by Faujas, Recherches ſur les Volcans, &c. p. 364. which he thinks burſt out of a granitic rock, and though in fuſion, inſtead of ſpreading and flowing horizontally, that it raiſed itſelf without any ſupport to the height of 100 feet; an event, which, as being contrary to

* See 5 Sauſſ. 6 Sauſſ. 127. 25 Roz. 174.

the

the nature of liquids, appears to me impoſſible; I ſhould rather ſuppoſe it a primeval trap, contemporary with granite, and depoſited by the fluid that contained both; in that ſpot the trappoſe matter was in greater plenty, and, therefore, raiſed to a greater height. In this manner the exiſtence of many trappoſe elevations in Le Forez, mentioned by Bournon in 35 Roz. may be explained. I ſhall only add, that the celebrated Mr. Charpentier, of Freyburg, having examined all the ſuppoſed volcanic products of Vivarois and Velay, judged them to be barely pſeudovolcanic. 4 Helv. Mag. 179.

In Italy the *Euganean* mountains have been judged volcanic by many; yet as, according to *Strange*, they conſiſt of, what he calls, lava and granite, few at preſent may be diſpoſed to embrace that opinion, unleſs ſuch as judge granite itſelf to be of igneous origin. In Germany, and particularly in Heſſia, Habichtswald and ſome other mountains were deemed volcanic by the late worthy, and highly ingenious, Mr. Raſpe, but ſome months before his death he told me, he had long given up that opi-

 nion.

nion. A volcanic origin has been aſcribed to many others, chiefly on the ground of their containing baſaltic pillars, but the neptunian origin of thoſe of Bohemia has been demonſtrated by Reuſs; of many of thoſe of Fulda by Karſten; and of moſt of thoſe near the Rhine by Noſe; and of ſome of them by Mr. Sauſſure; one of them, indeed, he ſeems to think volcanic. In all caſes where doubts may be entertained, whether a hill, or mountain, is volcanic, or neptunian, our judgment may, in my opinion, be governed by the following maxims:

1°. Where trap, or baſaltic columns, appear on, or form the body of the hill or mountain, of their uſual black, bluiſh, or greyiſh black, colour, there the hill or mountain may be deemed neptunian, at leaſt ſo far as concerns theſe; ſuch as are found on actual ignivomous mountains muſt have been thrown out with other neptunian ſtones, but in that caſe they are never erect, and, commonly bear ſome marks of heat.

2°. Where maſſes of ſhiſtoſe porphyry occur, of a greyiſh black, aſh grey, blackiſh blue, or greeniſh colour, and the felſpar appear

pear uninjured by heat, they, and the parts they repoſe on, are neptunian.

3°. Diſintegrated, or decayed, porphyries, or traps, wacken, and amygdaloids, may be diſtinguiſhed from indurated volcanic ſand and aſhes, piperino, pouzzolana, porous lava, reſpectively, by local circumſtances, and the changes which low degrees of heat produce in them, compared with the changes which the ſame variations of heat occaſion in the real volcanic products that reſemble them. Wacken containing mica can never be ambiguous. Beds of real volcanic aſhes, if ancient, are always interrupted or interceded by beds of earth, which ſome, without any proof, would have to be vegetable earth; and, if by this appellation, they mean no more than earth fit for vegetation, the appellation is juſt; but if they mean that ſuch earth was in all inſtances ſuch as had produced vegetables, they are certainly miſtaken, as Dolomieu has already noticed; this earth having been merely waſhed down by rain from the cinders and fragments of lava, with which it was originally mixed; wacken preſents no ſuch appearance; the ſtate of ſuch beds of volcanic

volcanic ſand and aſhes will be eaſily apprehended, by conſidering the ſtrata found in the promontory of Catanea, as exhibited by Count Borch in his Sicilian Mineralogy, p. 300.

		Feet.	Inches.
1	Porous black lava		10
2	Looſe earth		3
3	Porous, reddiſh, hard, lava		$8\frac{7}{12}$
4	Looſe earth		$4\frac{1}{12}$
5	Porous, reddiſh, hard, lava	1	$8\frac{2}{12}$
6	Looſe earth, full of aſhes		$3\frac{1}{12}$
7	Hard, black, lava, or indurated aſhes .		$10\frac{8}{12}$
8	Looſe earth, full of aſhes		$3\frac{1}{12}$
9	Hard lava, like No. 7	1	$8\frac{2}{12}$
10	Looſe earth		$4\frac{1}{12}$
11	Hard, black, porous, lava	1	$9\frac{2}{12}$
12	Looſe earth		3
13	Hard, porous, black, lava	1	$6\frac{3}{12}$
14	Black ſand	3	$3\frac{5}{12}$
15	More tranſparent ſand, whoſe depth was not diſcovered		
		14	1

Who does not ſee that ſuch thin beds of earth muſt have been formed in the manner I mentioned, and could never have been in a ſtate of vegetation?

4°. Volcanic ejections never preſent any undecom-

undecomposed pyrites; or at least these must be of posterior formation.

Again, whether a hill on which fire has evidently acted, has been a volcano, or a pseudo volcano, can be decided only by local circumstances, and the more or less partial effects of the fire it endured; this I cannot better illustrate, than by presenting the reader an account of the hill of Kamerberg in Bohemia, which Baron Born deemed an ancient volcano, but which Mr. Reuss, from local circumstances, has shewn to have been a pseudo volcano; in describing the external circumstances of this hill they both agree. It stands on a basis of micaceous clay slate (shistose mica), perfectly free from all connexion with the neighbouring hills: its length, from E. to W. is 1420 feet; its breadth, from N. to S. 720 feet, and it is one of the highest in the circle of Egra; its shape oblong conic; its stony masses lie bare on the S.W. side, they present rhomboids exactly fitted to each other; its summit, which is 95 feet higher than the surrounding field, discovers a cavity of 9 feet in depth, and 30 feet in diameter, overgrown with vegetation. About 60 fathom to the southwards, an excavation was practiſed to extract materials for a road, and in this the internal ſtructure of the hill may be ſeen: the ſtrata are nearly horizontal, and ſo much the looſer as they are nearer the ſurface; in denominating the foſſils, both authors vary; the denominations of each referring to the particular opinion of each, relatively to the origin of the appearances.

BORN.	REUSS.
No. 1. Black compact lava, with ſemitranſparent grains of ſhorl.	Greyiſh black baſalt, with olivins of various ſizes, fracture compact, uneven, ſp. gr. 2,96, unaltered by heat.
No. 2. Red, porous, ſpungy lava.	Reddiſh brown earthy ſlagg, with numerous minute pores.
No. 3. Black and yellow, porous, lava.	Browniſh black earthy ſlagg, with ſome minute ſcarcely viſible pores, and ſome large cellular, alternating; it contains alſo grains of quartz that have loſt their luſtre, but give fire with ſteel.
No. 4. The ſame as No. 2, but in minute fragments.	
No. 5. Compact, but ſomewhat	An earthy ſlagg, containing

BORN.	REUSS.
ſomewhat porous lava, containing fragments of grey ſlate.	taining bits of half burnt clay.
No. 6. Argillite, of which one ſide has its ſurface inlaid with white glaſs.	This, Reuſs could not find.
No. 7. Argillite, overlaid with black lava.	
No. 8. Grey pumice. No. 9. Unripe black pumice.	Neither of theſe could be found; the higheſt of the earthy ſlags had its ſpecific gravity 2,038.
No. 10. Red and yellow, unripe pouzzolana. No. 11. Yellow and brown clay, mixed with pouzzolana.	Fragments of earthy ſlags, more or leſs diſintegrated or withered.
No. 12. Greyiſh black, ſomewhat rifty, compact lava.	Trap or baſalt, ſomewhat altered by heat, being ſlightly porous, and rifty.

To prove this hill to have been only a pſeudo volcano, Reuſs ſtates, that ſtrata of coal lie at no great diſtance from it, and, probably, underlay it at a conſiderable depth. 2. That in various parts of Bohemia, theſe ſtrata of coal have been known to have taken fire, and to have produced appearances reſembling the foregoing. 3°. That on the S.W. part of the hill, baſalt or trap exiſts

 in

in beds, perfectly unaltered. 4°. That the alternations of earthy flags with half burned clay, might proceed from the alternations of marl, which is known to be easily fusible with clay. 5°. The supposed crater is much too inconsiderable, and may proceed from the contraction of the strata that had been torrified or ignited.

ESSAY VI.

OF THE INTERNAL ARRANGEMENT OF MOUNTAINS.

THE materials of which mountains consiſt, are diſpoſed either in irregular heaps, or piles variouſly interſected by rifts; or in beds or ſtrata ſeparated from each other by rifts, often horizontal, or varying from that direction by an angle of from 5 to 40 degrees, and ſometimes much more conſiderably, approaching even to a vertical poſition. The ſtrata of mountains are moſt frequently in the direction of their declivity, yet ſometimes their courſe is directly oppoſite, or *countercurrent*: the beſt manner of determining the angle of their courſe is by diſcovering that of their rifts, Charp. 80. it chiefly depends on the unevenneſs of the fundamental ground that ſupports them. Ibid. 57. According to 1 Sauſſ. 502, moſt of the elevated *granitic* mountains in Swiſſerland are formed of immenſe vertical pyramidal

pyramidal laminæ, parallel to each other, that is, piles ſomewhat inclining from the unequal diſtribution of their weight, a diſpoſition that may well be expected from *collateral* cryſtallizations; but this diſpoſition is not univerſal, for they have been found in Saxony horizontally ſtratified, Charp. 17 and 389. and in the Pyrenees, &c. ſee Eſſay V. § 1. much leſs can it be ſaid that this vertical poſition is general, for the ſtrata of *gneiſs* (which is only a modification of granite) are generally horizontal, Charp. 80, 81 and 191. and commonly very regular, diſcovering no traces of a violent ſhock. Ibid. 81. Mount Roſe, next to Mont Blanc, the higheſt in Europe, conſiſts alſo of gneiſs, which Mr. Sauſſure found horizontally ſtratified. 37 Roz. 17. 105. 8 Sauſſ. 113.

Shangin, who lately (1786) travelled over the Altaiſchan mountains, being conſulted by Pallas, whether he found any vertical layers or ſtrata therein, anſwered, he had not; but that he found them perfectly horizontal on the banks of the river Tſchary. 6 Nev. Nord. Beytr. 113.

Mountains of *primitive limeſtone* are frequently in irregular piles, but often alſo horizontally

horizontally ſtratified. Charp. 48. 216. *Siliceous ſhiſtus* is alſo often horizontally ſtratified. Charp. 22. Many *argillites*, particularly roof ſlates, are generally ſaid to have nearly a vertical poſition; but Voight has ſhewn that it is only their lamellæ that are ſo ſituated, their horizontal ſeams, and their walls, diſcovering their true poſition: their verticality ariſing only from the drain of the water, and, conſequently, their contraction in that direction; hence thoſe that are moſt ſilicited, as they contract leſs, diſcover leſs verticality: ſometimes horizontal ſtrata overlap on both ſides. 38 Roz. 289. 1 Sauſſ. § 447. Sometimes they are flanked on both ſides with vertical ſtrata. Ibid. § 339.

Much confuſion prevails in the ſtructure of the Pyrenees, and of the Grizon mountains, and thoſe on the border of the Baikal, and other great lakes, as may be ſeen in D'Arcet *Pyrenees*, 63. Ferb. Briefe Min. Inhalt. 11. Patrin, &c. from the cauſes mentioned in Eſſay III.

The perturbed ſtate of the ſtrata often proceeds from the decompoſition of internal beds of pyrites, to which water has had acceſs;

accefs; this appears to be the caufe of the alterations obferved in the mountain of Rabenberg, on the frontiers of Saxony. Charp. 253. In this mountain, a double direction of the ftrata of gneifs is obferved; between both the ftrata are vertical, and a large intermediate fpace is filled with iron ore: but this mountain contains beds of pyrites and vaft *fwallows*; moft probably then the pyrites fwoll, uplifted the whole, and the diffolved iron flowed into the vacuity, from which the water afterwards drained off on the fides.

In fecondary mountains, particularly the calcareous, the greateft diforder often prevails, though in general their ftratification is horizontal. Charp. 49. 39 Roz. 356. The vertical pofition of the ftrata fometimes proceeds from the fubverfion of fome of them, but fometimes alfo from calm depofition on primeval mountains, as I have hinted in my firft Effay, and is more fully explained by Mr. De la Metherie, in 42 Roz. from p. 300 to 306. Gruber alfo fhews this irregular pofition to proceed from internal avities, fo common in limeftone ſtone mountains, occaſioned moſt probably, by the eroſion of the interior by water. 2 Phy. Arbeit. 4, 5.

The calcareous mountains of Savoy are often arched like a *lambda*, probably from the ſinking of the intermediate ſtrata, the intermediate remaining horizontal. 1 Sauſſ. § 361. Sometimes they aſſume the form of the letters Z. S. C. or of a disjointed Ɔ C the convexities facing each other. Ibid, § 467, 475. So alſo in the Pyrenees, they ſometimes overlap, from an unequal diſtribution in their original formation. See Deſcrip. Pyr. 42. and bend various ways. Ibid, 72, and 102, and 162. where they aſſume a ſpiral form, or that of a horſeſhoe placed horizontally.

According to Lehman, moſt ſecondary ſtrata preſent hollows or *moulds*, (as they are called,) from internal depreſſion. Lehm. flotz. 137. But ſometimes alſo *elevations*, from an original elevation in the fundamental ſtone.

In Scotland, all the ſecondary ſtrata in the vicinity of primeval mountains, are nearly vertical; but at a greater diſtance

they

they approach more to an horizontal direction. 1 Bergm. Jour. 1789, 495, per Everſman.

The late ingenious Mr. Whitchurſt was led to imagine a greater regularity in the arrangement of ſtrata, than is really found in them: he tells us, p. 178, "that they "invariably follow each other as if it were in "an alphabetical order, or a ſeries of num-"bers, whatever be their denomination." —"not that they are alike in all the dif-"ferent regions of the earth, either in qua-"lity or in thickneſs, but that their order "in each particular part, however they "may differ in quality, yet they follow "each other in regular ſucceſſion, both as "to thickneſs and quality, inſomuch, that "by knowing the incumbent ſtratum, toge-"ther with the arrangement thereof in any "particular part of the earth, we may come "to a perfect knowledge of all the inferior "beds, ſo far as they have been previouſly "diſcovered, in the adjacent country." With reſpect to the ſtrata that accompany coal, he obſerves that "ſome inſtances are "apparently, but not really, contradictory "to this rule." Theſe obſervations are far

 from

from being univerſally true. For to ſay nothing of the coal mines in the valley of Plauen, in Saxony, where the ſtrata, though near each other, vary in thickneſs, from a few inches, to ſome feet, and that of coal, from 2 to 32 feet; nor of that in Mount Saleve, where the ſtrata of coal, though in a calcareous mountain, vary conſiderably, or of many others, he himſelf tells us that the lower ſtrata appear at the ſurface at Bonſal Moor, p. 188. And Mr. Pilkington, a later, and very exact writer, tells us, that in different ſiuations, the arrangement is not always the ſame. 1 Pilk. *Derbyſhire*, 55. And even when the arrangement is the ſame, the ſtrata differ in thickneſs in different parts of the country, p. 56. Again, Mr. Whitehurſt tells us, that wherever the ſtratum No. 1, a coarſe ſiliceous breccia or ſandſtone is found, No. 2, an indurated clay is found under it. And, wherever this indurated clay appears on the ſurface, No. 3, that is limeſtone, appears under it, p. 192. However true this may appear in Derbyſhire, though even there, this arrangement does not hold true in the coal mines, it is far

from

from being ſo in other countries, as is evident. See 2 Mem. Lauſ. 17, 2 part, and 52. He alſo obſerved that the ſtrata of limeſtone contain fiſſures filled with lead ore, but that theſe ſtrata are ſeparated by beds of toadſtone above 30 feet thick, into which, he ſays, neither the lead ore, nor any other mineral penetrates; yet the lead vein never fails of being found in the correſponding fiſſure of the inferior bed of limeſtone, p. 191, 193. Neither is this obſervation perfectly exact. Mr. Barker of Bakewell informed Mr. Werner, that the lead ore in many parts pierced through the toadſtone. Werner *Gange*, 139. Mr. Pilkington alſo tells us, that in the middle of it, lead veins 10 inches thick have been found, and in other places the lead has been found ſcattered through it. And that the vein in the inferior ſtratum of limeſtone, ſo far from correſponding, ſquints 4 or five yards from the ſuperior vein; into theſe errors Mr. Whitehurſt was betrayed by his fondneſs for the eruptive or plutonic theory. Moſt of his other general obſervations being grounded on this imaginary theory, are found to be deluſive,

or

or to hold good only in Derbyſhire and a few other countries.

The true reaſon why many of the ſtrata of England and Scotland appear ſo diſordered, is to be deduced from the ſhocks that Great Britain encountered at the period of the general deluge. This is not a mere hypothetical cauſe, but ſupported by undeniable proofs; and it is by a combination of the oppoſite impreſſions, from N. to S. and from S. to N. with the modifications which the obſtructions ariſing from the mountains of Scotland, and the intervention of Ireland between it and the Atlantic, that the various phenomena of its ſtrata may moſt ſecurely be explained; but ſuch an explanation can be attempted only after a minute and accurate examination of the whole iſland. Of the *dykes*, otherwiſe called *ſlips*, *troubles* or faults, I ſhall treat in the following eſſay.

The ſtrata of ſecondary mountains, generally, when not perturbed, aſcend towards the primitive. 2 Mem. Lauſ. 2 part, 25, in note, being formed upon them, or deſcending from them by diſintegration.

ESSAY VII.

ON COAL MINES.

MINERAL, or pit coal, is a ſubſtance whoſe external characters are too well known to require any ſpecification. Its weight with reſpect to water, extends from 1,23, to 1,500. The lighteſt is the beſt.

The object of this eſſay is to indicate the ſoils in which it is found, and the circumſtances attending them, thence to deduce a theory both of its origin and of the ſtate in which it now appears, together with the ſituations in which it moſt probably exiſts, and may be diſcovered.

CHAP. I.

§ 1.

Of Carboniferous Soils, and their attendant circumſtances.

By carboniferous ſoils, I mean the various ſorts of earth or ſtone among or under which coal is uſually found.

Thefe foils are either chiefly argillaceous, or arenilitic, or both together, or of the trap kind, or calcareous.

The circumftances of thefe and of the coal found among them moft worthy of notice, are the following.

1°. They commonly form diftinct ftrata, or beds one over the other to a great depth. The ftrata of coal are ufually called *feams*; it is very feldom found in irregular heaps, or in veins.

2°. Thefe feams are fcarce ever found *fingle*, but thofe whofe thicknefs does not exceed 14 or 15 inches, are rarely worked. At Whitehaven 5 were lately worked, at Newcaftle 3, at Liege 20. The higheft and next the furface are generally the worft, but the deepeft are not always the beft.

3°. The thicknefs of different beds of coal is variable, from half an inch or lefs to 5 or 6 feet; but not unfrequently it amounts to 25 or 30 feet, and in fome rare inftances, to 80 feet or more. No fuch feam as this laft has occurred in Great Britain.

4°. Seams of coal generally occupy a confiderable extent both in length and breadth, and whatever the thicknefs of each may be, it is commonly conftant for a confiderable ſpace, as a mile, or two miles; inſtances of a contrary kind ſeldom occur, unleſs the ſeam be diſturbed by ſome obſtruction, or at the extremities of a coal ſoil, or in an extent exceeding two miles*.

5°. In the ſame ſtratum, if exceeding 3 or 4 feet in thickneſs, the coal is ſeldom exactly of the ſame quality.

6°. Different ſeams of coal are ſeparated from each other, by at leaſt one, but generally by ſeveral ſtrata of earth or ſtone, as will be ſhewn in the ſucceeding ſections; theſe, in a conſiderable extent, preſerve alſo an uniform thickneſs.

7°. The uppermoſt ſeam of coal is commonly ſoft and duſty, it is vulgarly called *ſmut*.

8°. Seams of coal, and alſo their concomitant ſtrata, are generally parallel to each other, unleſs an uncommonly thick ſtratum of earth, 150 or 200 feet thick, intervenes †. Their number and order are alſo ſimilar, to a conſiderable extent, yet variable in the ſame diſtrict and ſoil.

9°. In many of the concomitant ſtrata,

* 1 Wms. 62, &c. 6 Lempe, 50. 2 Gerh. Betyr.

† Morand Arts and Metiers, Voly. 63. 1 Jars, 250. 6 Lempe, 50.

particularly

particularly of ſhale, bituminous ſhale, indurated clay, and ſandſtone, particles of coal are found interſperſed.

10°. The ſtrata that immediately cover coal, and thence called its *roof*, are ſhale, bituminous ſhale, or ſandſtone; rarely any other. But they are often alſo found at a great diſtance above it.

11°. The ſtrata on which coal repoſes, and thence called its *floor*, *ſole*, or *pavement*, are alſo ſandſtone, ſhale, indurated clay, or ſemiprotolite*. This laſt would, I believe, in moſt caſes, be found in its floor, if the mines were ſunk deep enough to reach it. Granite has alſo been found in its floor in a few inſtances. In trap ſoils, trap or baſalt is ſaid to form ſometimes the roof, and ſometimes the ſole of a ſeam of coal, but in ſtrictneſs, I believe, ſhale moſtly intervenes.

12°. Impreſſions of plants, particularly of the cryptogamia and culmiferous kind, are moſt frequently found on the ſhale and bituminous ſhale that accompany coal, or which are found in coal mines, ſometimes on ſandſtone, but very rarely on the coal it-

* 8 Buff. in 8vo, 232, 233. Semiprotolite is a reddiſh ſandſtone, or breccia, already mentioned.

 ſelf.

felf. Roots alfo frequently appear in the indurated clay. Trees carbonated, or bituminated fometimes repofe on coal, or are found under it. Fluviatile fhells, mufcles, and land fnails, often occur; fea fhells feldom.

13°. Argillaceous iron ore is fometimes met with among the carboniferous ftrata of an argillaceous foil; and martial pyrites either found or much oftener oxygenated, and mixed with the fubftance of the coal.

14°. The *ftretch* or courfe of feams of coal, and of their attendant ftrata, is commonly between E. and W. or N.E. and S.W. There are however a few exceptions to this rule.

15°. The *dip* of coal is exceeding variable, fometimes nearly horizontal, fometimes from 25 to 45 degrees, fometimes 75°, rarely approaching ftill more to the perpendicular.

16° The uniform courfe of feams of coal, and of the ftrata that accompany them, is frequently interrupted by obftructions called *flips*, *dykes*, *troubles*, *faults*, &c. Thefe never fail to el vate, or deprefs, the ftrata beyond them; or rather, the ftrata on each fide of them are found at different heights. This obfervation is general, being found to hold

hold good in every part of Britain, as well as on the Continent. The inequality of the height amounts from a few inches to 120 feet, but ſo great an inequality is rare, and has been ſound only in Derbyſhire. In Germany it ſeldom exceeds, and ſcarcely amounts to 50 feet.

17°. It has been obſerved in Britain that if the *ſlip* overhangs on one ſide, and conſequently forms an acute angle with the ſeam of coal which it cuts, the continuation of the ſtratum will be found *lower* on the other ſide of the ſlip, and conſequently *vice verſa* if it recedes from or forms an obtuſe angle with the ſeam of coal on the one ſide, the continuation of the ſeam will

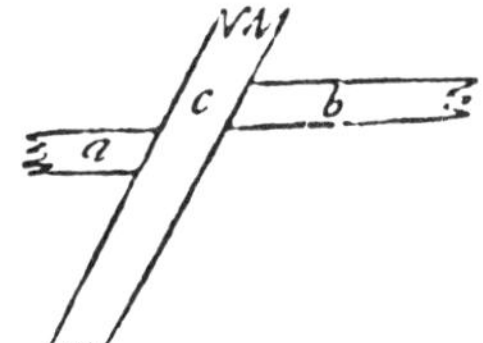

be found higher on the other, as in the figure, where *a* and *b* denote the interrupted ſeam of coal, and *c* the obſtruction or *ſlip**.

18° Theſe *ſlips* ſometimes conſiſt of indurated clay, ſometimes of ſandſtone, both different from ſuch as form the ſtrata, but more frequently of ſome ſpecies of

* 5 Ir. Acad. 275.

ſtone that never compoſe the ſtrata of coal mines, except, perhaps, rocks of the trap ſpecies; their thickneſs or extent amounts in various mines, from a few inches to ſeveral yards. Nodules of coal are ſometimes found among them, and water is frequently lodged in them. They often deſcend from the ſurface to the greateſt known depths.

19° The diſpoſition of the ſtrata below the ſurface ſeldom conforms to the figure of the ſurface. The former is often regular, when the latter is broken and uneven, and *vice verſa*, very frequently the ſtrata dip into a hill againſt the riſe of the ſurface, or croſs it in a right or diagonal line*.

20° The deepeſt mines known, are thoſe of the county of Namur, ſome of which are ſaid to deſcend 2400 feet.

21° The ſeams of coal where in contact with their roof, floor, or ſlip, have a ſmooth poliſhed gliſtening ſurface, Morand and 8 Buffon Mineralogy, 275, which ſhews they were originally ſoft.

* Wms. 114 Genſanne hiſt. Langued. 36, 37. 8 Buff. Mineralog. 8vo. 230, 231.

§ 2.

§ 2.

Of Argillaceous Soils.

Theſe conſiſt chiefly of *indurated clay*, which the miners commonly call *clunch*, and when much mixed with calx of iron, *bind*; or of *ſhale*, or *bituminous ſhale*; which miners, if I miſtake not, call *bat*; or of a decaying *porphyry*, which the miners call *rotten ſtone*; the argillaceous ſtone mentioned by Whitehurſt is, I believe, the ſame porphyry in a firmer ſtate; fragments of ſtone they call *ratchill*. If among ſeveral ſtrata of theſe, only one or two of ſandſtone occur, I ſtill call the ſoil argillaceous, and not *arenilitic*.

The ſoils about Whitehaven, and Newcaſtle, are of this ſort. At Newcaſtle the firſt ſtratum is a ſandy clay about 4 fathom in depth, ſucceeded by a brown ferruginous clay mixed with mica, as this is by a white micaceous ſandſtone, under which lies a bituminated clay mixed with mica, particles of coal, and pyrites, with calcareous incruſtation in the rifts, and under that lies the firſt ſeam of coal 6 inches thick, repoſing

repoſing on a thick ſtratum of indurated clay, the ſucceeding beds reſemble the foregoing. At Whitehaven, under a bed of common clay of 7 or 8 fathom, and 11 fathom of ſandy clay, a bed of natural clayey carbon or culm of 3 fathom is found, and under ſome beds of indurated, and ſome of micaceous clay, one of iron ore, one of ſandſtone, ſeams of good coal occur. 2 Ir. Acad. 163. At Alfreton Common, after a ſtratum of clay 7 feet thick, and another of ſtony fragments called rachill 9 feet thick, we meet with ſeveral beds of clay, more or leſs indurated, to the depth of 118 feet, and then find only *ſmut*. The bed of true coal is 40 feet lower. Whitehurſt's Theory, 211. here ſandſtone does not occur. At Weſt Hallam, the ſtrata are much the ſame, ſmut, the firſt indication of coal, is met with, but at the depth of about 56 feet, and under it clunch, with argillaceous globular iron ore, and roots running through the clunch. Pilkington, 81. and a bed of ſoft coal occurs about 53 feet lower. The beſt ſeam is found only at the depth of 214 feet. Whitehurſt, 212. To theſe I may add the coal mines at Hetruria,

truria, in Staffordſhire, as under the Ratchill, though a *limeſtone bed* only 1 foot thick occurs, the ſucceeding beds are ſand, and two beds of indurated clay, (then probably ſhale) and coal.

At Lichfield, to a bed of black clay 4 feet thick, and a bed of what is called rottenſtone 6 feet thick, there ſucceeds a bed of indurated clay, 18 feet thick (there called marl), and to this a bed of coal of 4 feet; under this lie other beds of indurated clay, one of which is *white*, of various thickneſs, and laſt, at the depth of 85 feet a bed of coal appears of the thickneſs of 30 feet. 2 Ir Acad. 100.

At Colebrook Dale, the ſtrata are remarkable; after 3 feet of brick clay and 15 of potter's clay, a ſeam of *ſmut*, 1 foot thick occurs; under this what is there called blue bat, 3 feet; then 7 feet of ſandſtone; then a ſeam of coal 4 feet; under this *white ſlip*, that is, potter's clay; under which is another bed of coal, ſucceeded by other beds of clay and coal, of various thickneſs. Ibid.

At Baldo, near Falkirk, in Scotland, after 7 feet ot clay, and 33 of ſlate (probably

bly ſhale), there occurs alſo a bed of *lime-ſtone* 3 feet thick, but it is ſucceeded by a bed of earth and ſhale 6 feet, which covers a bed of coal of from 2 to 6 feet. 6 Phil. Tranſ. abridg. Part II. p. 223. In the Princeſs's ſhaft, at Dolau, in Germany, the ſtrata are, grey calciferous clay, a fatter, not efferveſcent, a whitiſh micaceous ſlate, bluiſh ſlate, bluiſh ſandſtone, ſhale, a ſeam of coal, 2,5 feet thick, repoſing on bluiſh ſandſtone; with ſeveral alternations of the above with ſeams of coal; and, at laſt, ſemiprotolite. 2 Ger. Geſch. 156.

At Drim Gloſs, in Ireland, under 48 feet of clay and rubbleſtone, we meet a ſoft argillaceous ſtone 30 feet thick; then 35 feet of indurated clay, and 15 of ſhale covering 4,5 of coal. Whitehurſt, 246. Sometimes, but very rarely, coal is covered only by one ſtratum: thus at Ozegow, in Upper Sileſia, a ſeam of coal is found immediately under 18 or 24 feet of clay; the ſame circumſtance occurs in ſome mines in the county of La Mark, only with the interpoſition of ſhale. Georgi alſo obſerved an inſtance of this ſort in Dauria, near the river Argun and Schilca. 2 Gerh. Beytr. 159.

The

The ſtrata that occurred in ſinking a ſhaft at Ilkeſton, in Derbyſhire, are very numerous; and the thickneſs of thoſe of the ſame denomination, various; after 6 feet of ſoil and clay, there occurred 4 of ſhale; 18 inches of argillaceous iron ore; a *ſeam of coal* 15 inches thick; then clunch, grey ſtone, blue ſtone, black ſhale, iron ſtone, ſhale, blue bind, bituminous ſhale, blue clunch, blue bind, *coal*, 18 inches; black clunch, 4 inches; then a ſort of *cannel coal*, 9 inches thick; blue clunch, bind, light coloured ſtone 4 feet thick, a greyiſh blue ſtone as hard as flint, 6 feet, a very light coloured ſtone, 14 feet, bind, grey ſtone, blue bind, *ſoft coal*, 2,5 feet; bind, 6 feet, *ſoft coal*, 13 inches; black clunch, light coloured clunch, bind, 11 feet; coal, not an inch thick; black clunch 3 feet; clunch and bind, 25 feet; *hard coal*, 6 feet; clunch, 3 feet. Pilkington, 82.

At Pinxton church (near Alfreton, as appears by the map) the ſtrata are, 6 feet of ſoil and clay, black ſhale, with coal ſhreds; bind, ſtone, a crumbly ſtone, bind, hard ſhale, intermixed with ſtone, clunch with iron ore, bind, with ſtripes of coal, grey

ſhale,

ſhale, with ſhreds of black ſtone; yellow ſandſtone, ſandy ſhale, a ſiliceous ſtone, grey ſtone, with coal ſtripes, bind, 3 feet of ſmut; and 5 feet 10 inches of hard coal; the whole depth 77 feet. Pilkington, 86.

At Newhall, in Derbyſhire, the ſtrata are 12 feet of blue earth; 48 of black earth, then 3 feet of coal; then 3 feet of blue bind, covering 6 of hard coal; under this lie 3 feet of blue bind, covering 7 of a ſofter coal. Pilkington, 91.

In the valley of Plauen, in Saxony, ſeveral coal mines are found, whoſe ſtrata are thus deſcribed by Charp. 54. Under the mould, there occur clays of various colours, white, grey, reddiſh, or greeniſh; in ſome places 3 or 4 feet, in others 50 or 60 feet deep and there vulgarly called marl; 2°. then an arenaceous ſtratum, from a few inches to one foot thick, and of various colours. 3°. Grey, or bluiſh indurated clay, whoſe thickneſs is as variable as that of the firſt ſtratum. 4°. Another arenaceous ſtratum, as No. 2. 5°. A ſtill harder clay, of variable thickneſs, like No. 1. 6°. Shale, from 1 to 3 feet thick; and, laſtly, coal of different goodneſs, from 2 to 32 feet thick: this

coal

coal contains vitriolic acid, and in the ſtrata, vegetable impreſſions, and ſome of muſcles; bituminated wood has alſo been found in it, p. 58.

At Niederherſmdorf, in the ſame valley, coal has been found under a ſtony conglomeration of argillaceous porphyry, and argillite, 12 feet thick, but immediately covered by 36 feet of greeniſh and variegated ſhale; this coal abounds in pyrites. 2 Berg. Jour. 1792, 136, &c.

At Pottſchapel the ſtrata are, mould of various thickneſs, from 16 to 20 feet; then dark yellow, and greeniſh grey indurated clay, with impreſſions of plants, particularly of reeds, ſulphur pyrites, and fragments of galena, of variable thickneſs, from 40 to 120 feet, under which coal is found; ſometimes with the intervention of a ſtratum of bituminous ſhale 2 or 3 feet thick: the ſeam of coal is from 18 to 40 feet thick, 6 Lempe, 43, &c. It repoſes on indurated clay, as this does on a porphyritic rock. Ibid. 52.

In the coal mines near Bilin, in Bohemia, the ſtrata are, greyiſh mould, 2 feet; whitiſh grey clay, 5 feet; reddiſh brown ſand,

20 feet;

20 feet; bluifh grey clay, 10 feet; fhale with pyrites and vegetable impreffions, 18 feet; coal, 48 feet thick. Reufs Mittel Geb. 42. The direction of the coal is from S.W. to N.E. Ibid. 45, its dip is contrary to the declivity of the hill.

At the foot of the mountain of Gangelhof, which is partly porphyraceous and partly trappofe, there is a coal mine, the firft ftratum of which is mould, about 2 feet thick; the fecond, fand mixed with mica, 6,5 feet; the third, clay, 6,5 feet; the fourth, fhale; the fifth, coal, repofing on trap. Reufs Mittel Gebirg. 92. In the coal mines of Rive de Gier, the ftrata are, 1°. mould, 8 or 10 feet deep; then a farcilite with mica, 5 feet; another minuter grained; a coarfe fandftone, in fome places covered with fhale bearing vegetable impreffions; a micaceous fandftone; bituminous fhale; a feam of coal, 6 inches thick; fandftone, 10 feet; ditto flightly bituminated: ditto, bituminous fhale; fhale, bituminous fhale, coal, 18 feet. 13 Roz. 180.

§ 3.

§ 3.

Of Arenilitic, or Sandſtone Soils.

At Bagelt, in North Wales, after 45 feet of gravel and ſand, we meet with a bed of ſhale 9 feet thick; then argillaceous ſandſtone; and another bed of ſhale, under which a bed of coal is found; beneath theſe, indurated clay, ſandſtone, and coal, alternate with each other to a conſiderable depth; the ſandſtone in ſome ſtrata is 44, and in one 90 feet thick; in one inſtance a bed of ſhale, with ſhells of fiſh, covers the coal. Whitchurſt, 242.

The next mine of this ſort I ſhall mention, is that of Blakelow, near Macclesfield, which I owe to Mr. Mills; as he is well ſkilled in mineralogy, his deſcription is much more accurate than any of the foregoing, therefore I ſhall give it more at large.

The firſt ſtratum conſiſts of clay and gravel, 15 feet thick; this is followed by a bed of ſhale of the enormous depth of 102 feet; after which a thin ſeam of *ſmut* appears; under this a ſiliceous ſandſtone mixed with mica, 25 feet; then a ſoft grey lamellar ſandſtone mixed alſo with mica, about 2 feet, this is here called *grey beds*; then ſhale alternating with the grey beds to the depth of about 4 feet; under the laſt bed of ſhale another ſeam of coal is found, 14 inches thick, and under it a grey indurated clay; this is ſucceeded by a compact, ſiliceous, and ſomewhat micaceous ſandſtone, 36 feet thick; under which, after three alternations of grey beds and ſhale, a ſtratum of nodular iron ſtone occurs, 4 feet thick; then a bed of ſhale, covering another of coal, which repoſes on grey clay; this bed of coal *increaſes in thickneſs as it riſes in the hill.* The grey clay is ſucceeded by a conglomeration of ſhale and ſandſtone, 30 feet thick, under which lies a ſeam of coal ſo mixed with pyrites as not to be worked; the dip of this coal is to the N.E. nearly 26°. *Here we may obſerve,* that ſhale conſtantly covers coal, and that iron ſtone is found, and mica, and the beds as elſewhere, vary in thickneſs. 3 Ir. Acad. 50. In the ſhaft of St. Sophia, at Wettin, after 1 ½ foot of mould, there occur 5 fathom of a brown ſtone, and 6 of bituminous ſhale, which is ſucceeded by reddiſh blue ſomewhat micaceous, ſandſtone ſlate, as this is by a calcareous breccia filled with flints and whoſe cement is ſiliceous; this repoſes on a grey micaceous ſandſtone, which is followed by a blue limeſtone, under which lies a coarſe micaceous ſhale; then common ſhale covering a bed of coal, 2,5 feet thick, but mixed with pyrites; after this ſhale and coal alternate, until all terminate in red ſemiprotolite. 2 Gerh. Beytr. 152.

In the ſhaft of Dorothea at Lobegin, we meet the following ſtrata, mould, a pale yellow calciferous clay, yellow iron ſhot ſand, 6 fathom of greyiſh black ſlate (ſhale, as I ſuppoſe) ſomewhat micaceous, a grey breccia, with a ſiliceous cement formed of ſteatites, quartz and flints, 4,5 fathom thick, then ſhale covering a ſeam of coal about 4 feet thick, then a fine grained ſomewhat ſparry ſandſtone, another ſtratum of the above breccia, ſhale, coal, and ſandſtone. 2 Gerh. Beytr. 154.

At Ibenburen, in the county of Tuklenburg, after 3 or 4 feet of clay, we meet 6 feet of ſlaty micaceous ſandſtone, and under it a thin bed of coal, then a bed of ſhale, one of coal, and three of ſandſtone; laſtly a bed of ſhale covering another of good coal, 8 feet thick. 2 Gerh. Geſch. 157. In the coal mine of Boſerup, in Scania, the firſt ſtratum is mould ſomewhat clayey and ſandy 12 feet, a grey argillaceous ſandſtone with iron ore, 20 feet, ſucceeds; then a bed of coal only 1 foot, repoſing on indurated martial clay, then a yellowiſh ſandſtone, 7 feet, a ſandy ſhale 3 feet, covering a pyritous coal 2 feet thick. Schwed. Abhand. 1773. The mine at Helſingham is ſomewhat ſingular; below a bed of ſandſtone 36 feet thick, appears a ſeam of coal, 2 feet, then ſhale 12 feet, ſandſtone 6 feet, and, laſtly, coal 1 foot. Ibid.

In the coal pits of Shubley, in the pariſh of Dronfield, in Derbyſhire, under 90 feet of ſandſtone there are 33 of blue bind, and 21 of black ſhale, covering 3 of coal, under which is a bad ſort, called dirt, and again, 2 feet of coal. Pilkington, 87. In the coal pits of Morſleben, and Weſenſleben, in the principality of Halberſtadt, the following ſtrata occur, mould, of variable thickneſs, a ferruginous ſandy clay, 6 feet, grey clay, from 18 to 24 feet, brown ferruginous ſandſtone,

ſtone, whitiſh grey ſandſtone, bluiſh ſandſtone, a bluiſh ſlaty and ſandy argillocalcite, indurated ſlaty clay, a bluiſh ſlaty argillocalcite, coal, from 10 to 18 inches thick (Gerhard, 105), ſhale, 6 or 8 feet, black ſlaty, indurated, clay, grey argillaceous ſandſtone, from 24 to 30 feet thick, ſlaty indurated clay, grey ſandy argillocalcite, with pyrites intermixed. Lehm. 184. At Althal in the foreſt of Thuringia, a ſeam of coal repoſes on white ſandſtone, but ſometimes immediately on granite. 2 Berg. Jour. 1790, 322. At Themmin in Bohemia, the ſtrata are, mould, turfy earth, ferruginous ſandſtone, marl, coarſe ſandſtone, ſhale, coal, repoſing on ſandſtone, 2 Berg. Jour. 1791, 57.

Near Mannebach, coal is found in a rift of a porphyritic mountain covered with ſhale, and floored with ſandſtone, in ſeveral ſucceſſive ſtrata. Voight, 9, 10. In the neighbourhood of Liege there are upwards of 40 ſeams of coal one above the other; they are ſeparated from each other by coarſe, or fine grained, ſandſtone, frequently micaceous, and ſometimes by black reddiſh ſhale. 1 Jars. 288. 292. the uppermoſt ſeams are

often

often the beſt; in the ſame ſtratum the coal is often of different qualities in different diſtricts. At Aix la Chapelle, the coal alſo lies between ſtrata of ſandſtone, and ſhale or indurated clay. It is moſtly natural carbon. 1 Jars. 306.

Of Trappoſe, or baſaltic Soils.

At Bally Caſtle, 60 feet of trap or whin, cover 24 of ſhale; under which a yellow ſandſtone 42 feet, covers 21 of coal; this reſts on 90 feet of grey ſandſtone, which cover another ſeam of coal 5 feet thick. Whitehurſt. At Borrowſtounneſs in Scotland, a ſtratum of trap or whin is the immediate roof of a ſeam of coal; and at Hillhouſe near Lithlingow, a thin ſeam of coal is found below a ſtratum of columnar baſalt. At Bathgate hills, ſtrata of coal and baſalt alternate with each other. 1 Williams, 70.

Carbonated wood is frequently found under trap, whin, or baſalt.

At Stackhouſe, in Weſterwald, after 6 feet of mould, we meet 6 of whin or wacken; then ſome ſtrata of brown or grey clay; under which the firſt ſtratum of carbonated

bonated wood appears; ſometimes the firſt ſtratum is black, being mixed with whin or wacken; the 2d a grey clay; the 3d a yellow clay: and then a bed of whin or wacken; but the coal is always covered by blue or grey clay, its depth from 11 to 14 fathom. 1 Berl. Beob. 52. 56. Neither petrifactions nor vegetable impreſſions are found in it, 58. but cryſtals of ſelenite and pyrites abound. The thickneſs of the ſtratum continually decreaſes in receding from Stackhauſen, 106. 108. Wood coal and bituminous ſtone coal frequently accompany each other; ſometimes, as at Meiſen, the ſtone coal is uppermoſt, and the wood coal under it; but at Toplitz, in Bohemia, the wood coal is uppermoſt. 110.

At Meiſen in Heſſia, galleries have been puſhed upwards into the ſide of the hill, 900 feet under its ſummit. 1°. Into ſandſtone, then into clay, and, laſtly, into a bed of coal from 6 to 90 feet thick, over this lies a maſs of trap or baſalt, 600 feet high: the courſe of the coal is horizontal, and the baſalt in ſome places penetrates down to the clay; where the coal is ſome fathom thick it forms a ſtrata, that next the baſalt is the best and most bituminous. 1 Berg. Jour. 1792, 379. Under that is a slighter kind reposing upon wood coal; under which is a stratum, ill described, but which I take to be a mixture of disintegrated coal and clay. 1 Berg. Jour. 1789, 275. 280. Between Mezin and Velay, natural carbon is found under trap or basalt, and reposing on granite. 2 Soulavie, 235. 237. yet Faujas 8 Buff. Mineral. in 8vo. tells us, he always observed a bed of semiprotolite interceding between the coal and the granite; Faujas also observed basaltic pillars reposing on a seam of coal, and not piercing through it; yet there interceded between them a thin bed of grey clay; the stratum of coal was 3 inches thick, it rested on another of grey clay, under which was another thick seam of coal. Faujas Recherches sur les Volcans Eteints, 388. The beds of clay, he observed, proceeded without interruption into the mountain, which shews they did not proceed from the decomposition of the basalt. Ibid. 339.

Of

Of calcareous Soils.

Near Milhant and St. George, coal is found in a ſecondary calcareous mountain, but covered by bituminous and pyritiferous clay. 11 Ann. Chy. 272. Near Kratigen in Swiſſerland, a thick bed of coal is found in the midſt of limeſtone. Ferber Briefe Mineralog. Inhalt. 24.

At Multhorp in Sweden, a coal reſembling cannel coal is found incloſed in ſtrata of aluminous ſhale, over which 2 ſtrata of limeſtone repoſe. Mem. Stock. 1767, 35. At Alais, the coal is ſo mixed with limeſtone as to afford lime after combuſtion. 8 Buff. Mineral. 8vo. 189.

Between the calcareous ſtrata of Mount Saleve in Swiſſerland, there are thin ſeams of coal, roofed and floored with grey or brown calciferous ſhale. 1 Sauſſ. § 246. This coal ſeems to be of the nature of cannel coal. Ibid. At Lobegin near Wettin, the ſtrata are, 1. Mould of various thickneſs, then loam 6 feet thick, red ſand, 7 feet, ſlaty argillaceous ſtone, 10 feet, grey ſwineſtone, 10 feet, grey limeſtone, alternating with an iron ſhotſtone, both con-

taining,

taining, the firſt white, the other red baroſelenite, from 12 to 18 feet, a grey argillocalcite of variable thickneſs, ſhale with ſhreds of coal, black indurated clay with neſts of calcareous earth and ſulphur pyrites, coal, ſelenitic ſpar (baroſelenite), coal, 3,5 feet, under which lies a bad kind, grey limeſtone 3 or 4 feet, black ſhale, and limeſtone, more or leſs argillaceous. Lehm. 180.

Strata of limeſtone, of various thickneſs, colour, and quality, are very common in the coal fields of Scotland; at Blackburn, in Weſt Lothian, a ſtratum of limeſtone, 6 or 7 feet thick, is the immediate cover of a ſeam of *caking coal* 5 or 6 feet thick. At Carlops a ſtratum of coal is found beneath a limeſtone quarry. 1 Williams, 73. 1 Berg. Jour. 1789, 495. The limeſtone of Coal fields contains a variety of ſhells, coral, and other marine productions, blended in the heart and compoſition of the ſtone. 2 Williams, 26. At Anzin, near Valenciennes, there is a tract in which horizontal beds of limeſtone, marlite, and chalk, alternate, under theſe are ſandſtone and ſhale, between which coals lie. 11 Ann. Chy. 273. In

ſome

many of the ſolid ſubſtances which form the great maſs of the globe, it is equally evident that it muſt be coeval with thoſe ſubſtances, and, conſequently, muſt have preceded the exiſtence of animals and vegetables, as I have ſtated in my firſt Eſſay.

Before this important diſcovery, three ſolutions of this problem were propoſed, two of which have ſtill ſome footing, and the third being almoſt generally received, I muſt previouſly ſhew the inſufficiency of all of them, as an apology for introducing my own.

According to the firſt opinion, which is that of Mr. Genſanne and ſome other mineralogiſts, pit coal is nothing elſe than an earth or ſtone, chiefly of the argillaceous genus, penetrated and impregnated with petrol or aſphalt. Kilkenny coal or natural carbon demonſtrates the inſufficiency of this ſolution, for this coal contains neither petrol nor other bitumen; beſides, the quantity of earthy or ſtony matter in the moſt bituminous coal, bears no proportion to the weight of that coal; bituminous coal is capable of being charred, and then it is a ſubſtance almoſt entirely reſembling

vegetable

vegetable charcoal, which, on combuſtion, ſcarcely leaves $\frac{1}{30}$ of its weight of argil or ſtony matter: neither can any be ſaid to be a bitumen, for when charred, its volume is diminiſhed at moſt $\frac{2}{3}$, whereas bitumen, even the moſt compact, either leaves no coaly matter, or at moſt, only an inconſiderable quantity.

The next, and ſtill the moſt prevailing opinion is, that all mineral coal is of vegetable origin, that it ariſes from the immenſe foreſts with which the earth was originally covered, which by various ſubſequent revolutions were buried under thoſe vaſt ſtrata of earth which at preſent cover mineral coal; theſe woods, it is ſaid, were mineralized by ſome unknown proceſs, but of which the vitriolic acid was the principal agent; by means of this acid the oils of the different ſpecies of wood were converted into bitumen, and a coaly ſubſtance was formed, as in the proceſs for making æther; in ſupport of this opinion it is alledged, 1. that in many of the various ſtrata that cover coal mines, and particularly in the ſhale or indurated clay, which immediately cover ſeams of coal, vegetable impreſſions

preſſions are found. 2. That wood, actually converted into coal, is frequently obſerved among ſtrata of mineral coal, and, in ſome, whole trees are found, ſome parts of which are in their original vegetable ſtate, and the remainder converted into coal.

I ſhall firſt obviate the conſequence deduced from theſe obſervations, and then ſtate the principal objections to this ſyſtem.

That vegetable impreſſions are often found in the ſtrata that cover coal, or on thoſe on which it repoſes, is a certain fact, but this fact may be better accounted for in the theory I ſhall preſently propoſe; the impreſſions obſerved are thoſe of herbaceous plants, as ferns, &c. and theſe, of all others, contain leaſt oil; their exiſtence has, therefore, no connexion with the converſion of vegetables into coal; the impreſſions of reſiniferous plants, which alone are capable of furniſhing moſt oil, have never been diſcovered on the ſtrata that accompany coal, and the trees found are commonly birch or oak. As to the agency of the vitriolic acid, it is very different from that which has been ſtated, no vegetable oil has ever been converted by it into petrol, nor has

has it ever been known to have contributed to the formation of any real bitumen; its agency however, in converting ſome ſpecies of wood into coal, when bitumen is otherwiſe ſupplied, is not denied; it acts upon, decompoſes, and is decompoſed, by the reſinous part of the wood, or its oil, and leaves the carbonaceous part untouched; but this carbonaceous part would never be bituminated and converted into coal if real bitumen were not preſent, as appears by the obſervation of Arduino, who found the timber employed in ſupporting ancient mines, and which had ſtood many centuries in the midſt of pyrites and vitriolic waters, blackened indeed, but not in the leaſt bituminated; whereas wood anciently depoſited in the muddy bed of the Lagune, about Venice, was in ſome degree bituminated, having received bitumen from the ſea water. Fraboni del Antracite, 11. In this caſe the wood was firſt decompoſed by putrefaction, and afterwards ſomewhat bituminated. Hence it is allowed that ſome fragments of wood may have its texture, in great meaſure deſtroyed by the vitriolic acid, and be afterwards bituminated; yet its

its texture is never totally destroyed, but may be discerned by the help of the nitrous acid, as Mr. Fraboni has shewn *.

Thus that species of coal called wood coal is formed, in which vitriolic acid is always found, but it is easily distinguished from true mineral coal, as it burns more weakly, with a disagreeable smell and abounds in vitriolic acid; whereas true mineral coal contains this acid only accidentally, emitting most frequently no acid in distillation, but only volalkali, as Mr. Lavoisier has shewn. Mem. Par. 1778, 436. The ashes of wood coal afford some traces of fixed alkali †, whereas true mineral coal contains none, at least Model found none in that of Newcastle. 1 Model, 456. This shews that mineral coal does not originate from wood deposited in or out of the sea:—Yet as it may be replied, that nature probably possesses means of mineralizing vegetables, with which we are unacquainted, perhaps the following ob-

* See my Mineralogy, vol. 2, p. 61.

† 2 Gerh. Beytr. 271. Fabroni, 58. Venel. sur la Houille, 37.

ſervations may be deemed the moſt conclusive.

1°. Coals are commonly found in ſtratified mountains, and form ſtrata, of which, however, they may differ one from the other in thickneſs, each preſerves its *own* for a conſiderable ſpace, as half a mile, a mile, or two miles; but in mines of *wood coal* no ſuch uniformity takes place; on the contrary, in the moſt conſiderable of theſe an uniform decreaſe of thickneſs from the place in which the wood was firſt heaped, is obſerved; thus in the famous mine of wood coal, in the Weſterwalds in the territory of Orange Naſſau, the thickneſs of the ſtratum is obſerved continually to decreaſe as it extends from Stockhauſen. 1 Berl. Beob. 106. 108. and Morand, in Deſcription des Arts and Metiers, p. 10, folio.

2°. Seams of real mineral coal, and thoſe of earth or ſtone that accompany them, are obſerved, while uninterrupted by ſlips or dykes, to lie parallel to each other to a great extent, and even after ſuch interruption, whether elevated or depreſſed, ſtill to maintain their parallelism. But in mines of wood coal, notwithſtanding the contrary

 aſſertion

aſſertion of Morand, Ibid. p. 8, 9. late obſervations have aſcertained that no ſuch parallelifm, nor even any diſtinct number of ſtrata prevail, but the whole appears to be one ſtratum irregularly divided by maſſes of clay or ſtone; this has been proved in the Weſterwalds (the very place in which Morand, miſled by ſuperficial obſervers, had aſſerted a regular ſtratification to have taken place,) after an accurate examination by Mr. Becker ſecretary to the Mining Board of that country, 1 Berl. Beob. 101. 103. and by Fabri in his account of the mine of wood coal at Bruchliz. 1 Lempe Mag. 142.

3°. Mines of wood coal preſent ſudden elevations or depreſſions in the ſame ſtratum; mines of real mineral coal never. 1 Berl. Beob. 58.

4°. There are no ſlips or dykes in wood coal mines, 1 Berl. Beob. 58. thoſe of genuine coal abound in them.

5. Wood coal is frequently covered with round fragments of quartz, 1 Lempe. Mag. 143. genuine coal never.

6°. There is at preſent in the Muſeum of Florence, a cellular ſandſtone, the cells of which are filled witn genuine mineral coal.

6 Fabroni

Fabroni Dal Antracite, 10. Could this have been originally wood?

7°. According to Voight Prack. Geb. 80, 81. genuine coal is ſeldom found in plains, but wood coal frequently is. Ibid.

Reflecting on theſe facts, it appears to me highly probable, that real mineral coal does not originate from vegetable ſubſtances of any ſort; the reſemblance obſerved between bituminated carbonated wood and mineral coal, ariſes from the ſimilarity of their *compoſition*, both being formed of carbon and bitumen, but by no means evinces the *filiation* of the latter from the former.

The third opinion relative to the origin of pit coal is, that of the celebrated Arduino; he thinks it entirely of marine formation, originating from the fat and unctuoſity of the numerous tribes of animals that inhabit the ocean. Fabroni, 12. 14. This opinion reſts ſimply on the obſervation, that ſea ſhells, or their impreſſions, are frequently found in coal, in Italy, or at leaſt in, or on, the ſtrata that accompany them, but it is fully contradicted by the much more general obſervation, that the traces of land and not of marine vegetables are found on the ſtrata that cover ſeams of coal in all countries, or on thoſe on which theſe ſeams reſt, or on both: that ſea ſhells are ſcarce ever found among them in other countries, and much leſs the bones of fiſh: that, on the contrary, reeds or ruſhes, and fluviatile ſhells, have been found in the ſtrata that cover coal. Morand, 8. That common ſalt is never found in coal mines, except when in the neighbourhood of ſalt ſprings, 2 Gerh. Geſch. 144. but, on the contrary, alum and vitriol, which are never found in the ſea. 3 Mem. Lauſ. 321. That in the deſerts near the Caſpian, on which the ſea is known to have reſted for perhaps many centuries, no coal mines are found: that the unctuoſity of marine animals ſhould rather float than ſink in the ſea;—that there is no known inſtance of its having been ever converted into a bitumen.

The opinion which I now propoſe, is grounded on the following facts:

1°. *Granite* has been known ſometimes to contain ſmall veins of coal. 30 Roz. 378. per Bournon, and Lavoiſier in Mem. Par. 1778, 440.

2°. *Plombago*, which is natural carbon combined

combined with a certain proportion of iron, has alſo been found in granite, 2 Berg. Jour. 1790, 532. and in *ſhiſtoſe mica.* 2 Sauſſ. 450, 451.

3°. *Hornblende,* a ſtone which enters into the compoſition of many rocks of the granitic order, as ſienite, and grunſtein, and of many granitines, and of moſt traps, and of many baſalts, has lately been found to contain carbon. Lampadius, 184. Hornblende is alſo often found in gneiſs, and ſtill oftener in porphyry; carbon has alſo been diſcovered in *ſiliceous ſhiſtus* by Wiegleb, 1 Chy. Ann. 1788, p. 50 and 140. and in baſanite by Lampadius and Humbolt. 2 Chy. Ann. 1795, 114 and 3.

Again, bitumen alſo is found in various ſtones, and flows from various mountains: in Mount Caucaſus there is a fountain of it that flows into the ſea, and ſinks to the bottom. 3 Deſcouvertes Ruſſes, 83, 84. Mr. Mills diſcovered a bitumen whoſe ſpecific gravity was 1,284, immediately under a *trap* (he calls it lava). Phil. Tranſ. 1790, 83. 87. It has been found in quartz, 1 Berg. Jour. 1791, 91. in the cavities of trap in Derbyſhire, 1 Pilk. 178. and in black marble. Ibid. 178. Fortis alſo obſerved it in black marble in the Iſle of Bua, on the coaſt of Dalmatia, tranſuding in the heat of the day, and congealed into drops at night*. Williams made the ſame obſervation in Scotland, 1 Williams, 235. and Triewald in Sweden. Swenſk. Handling. 1740, 203. Its exiſtence in bituminous marlite, and bituminous ſhale, is well known. Pallas diſcovered bitumen in the limeſtone on the banks of the Volga; 1 Deſcouv. Ruſſes, 462. and in Bavaria there are fountains of it in mountains conſiſting of ſandſtone and limeſtone, that alternate with each other. Flurl, 89.

A whole lake of aſphalt is ſaid to exiſt in the Iſle of Trinidad, in South America, which is liquid in ſummer, and ſolid in winter. Phil. Tranſ. 1789, 65. And ſeveral in Siberia, 1 Deſcouv. Ruſſes, 384, &c. and Perſia.

Fountains of bitumen have been found in coal mines. Mem. Par. 1747, 1031, in 8vo.

Hence it evidently follows, that carbonic

* Travels to Dalmatia, Eng. Edit. in 4to. p. 177.

ſubſtance,

ſubſtance, and petrol entered into the original compoſition of the ſtones already enumerated, and therefore are derived from the primordial chaotic fluid, in whoſe boſom moſt ſtones were formed.

I muſt farther obſerve, that it is admitted that moſt mountains were originally much higher than at preſent, having been ſucceſſively degraded and lowered by various ſubſequent accidents; particularly by diſintegration and decompoſition. That many of the granitic and porphyritic order, particularly thoſe in which hornblende is found, and thoſe of the trappoſe order, and ſiliceous ſhiſtus, are moſt ſubject to this degradation and decompoſition, and thoſe of the calcareous and argilitic genera, are leaſt ſubject to it. It is alſo known, that the higheſt mountains, from the greater viciſſitudes of heat and cold, and the impetuous ſhocks of the atmoſphere, to which (every thing elſe being equal) they are moſt expoſed, are moſt ſubject to this deſtructive operation,

That ſeveral mountains have been entirely deſtroyed by decompoſition.

That the higheſt mountains condenſe moſt

 water,

water, which gradually trickling down their ſides, carries off moſt of their diſintegrated parts.

From theſe facts it may juſtly be inferred,

1°. That natural carbon was originally contained in many mountains of the granitic and porphyritic order; and alſo in ſiliceous ſhiſtus, and might, by diſintegration and decompoſition, be ſeparated from the ſtony particles.

2°. That both petrol and carbon are often contained in *trap*, ſince hornblende very frequently enters into its compoſition.

My opinion, therefore, is, that coal mines or ſtrata of coal, as well as the mountains or hills in which they are found, owe their origin to the diſintegration and decompoſition of primeval mountains, either now totally deſtroyed, or whoſe height and bulk, in conſequence of ſuch diſintegration, are now conſiderably leſſened. And that theſe rocks anciently deſtroyed, contained moſt probably, a far larger proportion of carbon and petrol, than thoſe of the ſame denomination now contain, ſince their diſintegration took place at ſo early a period.

On

On this ſuppoſition, I think the formation of coal mines, and moſt of the circumſtances attending them, may naturally be accounted for.

And firſt, as to the ſeams of coals themſelves and their attendant ſtrata, they muſt have reſulted from the equable diffuſion of the diſintegrated particles of the primitive mountains, ſucceſſively carried down by the gentle trickling of the numerous rills that flowed from thoſe mountains, and, in many caſes, more widely diffuſed by more copious ſtreams. By this decompoſition the felſpar and hornblende were converted into clay, the bituminous particles thus ſet free, reunited and were abſorbed partly by the argil, but chiefly by the carbonaceous matter with which they have evidently the greateſt affinity, ſince they are ſeparable by boiling water from the former, and ſcarcely by the ſtrongeſt heat in cloſe veſſels from the latter, and even in an open fire, only by a heat much ſuperior to that of boiling water. The carbonic and bituminous particles thus united, being difficultly miſcible with water, and ſpecifically heavier, ſunk through the

the moiſt, pulpy, incoherent, argillaceous maſſes, and formed the loweſt ſtratum, unleſs in caſes where their proportion to the argillaceous particles was ſo ſmall, that the latter had ſubſided and coaleſced before the former could have been reunited, in that caſe the clayey particles formed the lower ſtratum of indurated clay. But if the petrol were in the greateſt proportion, it freqnently ſunk firſt in the form of a ſoft bitumen, carrying with it the clay and forming beds of ſhale, or bituminous ſhale, according to its proportion. By oxygenation it becomes ſpecifically heavier than water.

In ſome caſes, ſemiprotolite* formed the lower ſtratum. This happened when ferruginous particles were moſt abundant and ſeparated before the diſintegration, or at leaſt, the decompoſition of the ſtony particles could be effected.

In other obvious caſes, porphyry ſeems to form the lower ſtratum, namely, where primitive porphyry formed the ſubſtance of the plains over which the diſintegrated particles were diffuſed; and this ſeems fre-

* See 1 Eſſay, p. 44.

quently

quently the caſe in Saxony and England. It is only at ſtated periods of ſucceſſive years that the diſintegrated particles thus diffuſed, formed maſſes ſufficiently conſiderable to allow the liberation of a ſufficient quantity of carbonaceous and bituminous particles to form diſtinct ſtrata of coal and earthy matter; from the regularity of theſe periods, the regularity of the ſimilar compoſition of the diffuſed particles, and the uniformity of their diffuſion, the regular thickneſs and paralleliſm of the ſtrata naturally originated.

As theſe ſtrata were formed, not under the ſea, but long after the emerſion of their parent mountains from its boſom, hence it it is that marine ſhells, or the impreſſion of marine vegetables, are ſcarce ever found in them, Mem. Par. 1747, 1073. not even in thoſe of Whitchaven, which dip and are worked under it; but, on the contrary, trees and fluviatile ſhells are frequently found among the ſtrata; thoſe of Whitehaven ſtretch under the ſea, becauſe the ſea covers what before the inſular ſtate of Britain was land; but in Italy, and other parts contiguous to the Mediterranean and Adriatic,

Adriatic, marine remains more frequently occur, from the causes mentioned in Essay III. The herbs whose impressions very generally occur on the shale that immediately covers coal are, as Morand observes, those that grow on low and moist grounds*, as some species of the cryptogamia. It is true that Jussieu, Mem. Par. 1718, and 42 Roz. 450, thought these herbs to be peculiar to the East and West Indies; but Mr. Blumenbach, in this more advanced stage of the botanic science, judges them to be perfectly unknown†, as indeed should be expected; such soils so composed and so situated being scarcely to be met with, and if they were, that species may at this time be easily suppressed by other more numerous tribes: some, however, much resemble the ferns that now exist in the adjacent country. Mem. Par. 1747, 1033. The impressions of these herbs is easily accounted for from the soft oleaginous state of the shales on which they are found, which readily received their print, and retained it

* Morand *Art* du Charbonier, in Descript. des Arts, &c. p. 170.

† Hanbuch der Natur. Gesch. 703.

long

long after the originals had decayed. This circumſtance fully proves the ſoft original ſtate of theſe ſtones, and that the ſtrata had experienced no violent concuſſion ever ſince their formation, which dates from the remoteſt antiquity, as the ſlighteſt friction would irretrievably efface theſe impreſſions, and, conſequently, that the numerous revolutions, and eruptions of lavas from ſubterraneous fires, which many theoriſts ſuppoſe to have taken place, are purely viſionary. Nevertheleſs, that ſome derangement of the ſtrata, but gentle and gradual, has taken place ſince their formation, is evident from their diſlocation on different ſides of, what is called, a *ſlip*, *dyke*, or *trouble*. The effect of this obſtruction we have already mentioned; when it leans on one ſide, for inſtance to the eaſt, and, conſequently, overhangs the ſtrata of coal placed to the eaſtwards, the continuation of the ſtrata of coal, and their concomitants, will often be found *lower* on the weſtern ſide.

To underſtand the reaſon of this phenomenon we muſt remember that where it takes place with the circumſtances juſt mentioned, the *ſlip* is always a ſubſtance more or leſs different from any that form the ſtrata

ſtrata in the coal mine, and when a ſtone, that it is a ſtony maſs that reſiſts decompoſition, whilſt the maſſes that formed the ſtrata underwent that operation.

2°. That no traces of fracture or compreſſion of the ſtrata are found at conſiderable diſtances from it, and, conſequently, that it did not roll down from the diſintegrated mountain from which theſe ſtrata originate, at any time after the formation of the ſtrata, otherwiſe, from its great weight, ſome traces of oppreſſion would be obſerved; conſequently, it exiſted in the ſpot where it ſtands, before the ſtrata were formed, and as its height is ſo conſiderable as to reach through all the ſtrata, it muſt have obſtructed their extenſion for a conſiderable time, though at laſt the ſtreaming waters muſt have conveyed the diſintegrated particles beyond it; the preſſure therefore on the ſide on which the ſtrata were firſt formed, muſt have been much more conſiderable than on the ſide on which the ſtrata of later formation repoſe, and muſt have puſhed the upper and moſt moveable extremity of the *ſlip*, gradually towards the ſide on which there was leaſt preſſure and reſiſtance; on that ſide it muſt therefore overhang;

hang: this preſſure being of earlier date than that on the oppoſite ſide, muſt have had a more conſiderable effect in depreſſing each particular ſtratum, by ſqueezing out the watry particles entangled in them, and forcing their integrant particles into cloſer contact than could have been produced (the times being unequal) in thoſe of later formation, and, conſequently, the ſtrata muſt be lower. This explanation, I own, I am not perfectly ſatisfied with, and only acquieſce in, until a fuller account of the circumſtances of theſe ſlips is obtained. Where ſuch a connexion between the *inclination* of the ſlip, and the elevation or depreſſion of the ſtrata has not been obſerved, and I do not find that it has in France, Flanders, or Germany *, yet the unequal poſition of the ſtrata on each ſide of it conſtantly has, and is attributed by Mr. Charpentier, with great ſhew of reaſon, partly to the unequal elevation of the baſis on different ſides of the ſlip, and, in ſome caſes, to the ſliding away of ſome of the lower ſtrata. *Saxony*, 373. In Vol. VI. of the Phil. Tranſ. abridg. Part II.

* Charp. 373, Morand Deſcription des Arts et Metiers, in folio, Charbonier, 58. 1 Jars. 291.

p. 185.

p. 185. the ſituation of metallic ſtrata diſlocated by ſuch obſtructions is repreſented as the very reverſe of that which takes place in coal mines, and its explanation is much more natural.

That the cauſes aſſigned by Charpentier, muſt have operated in ſome caſes, is evident from this, that a ſimilar diſlocation of the ſtrata has been obſerved where the intervening ſlip was only a few inches thick, as at Zwickhaw. 6 Lempe, 55. and no diſlocation at all, or ſcarce any, where the ſlip was of conſiderable extenſion. Ibid. In the firſt caſe, therefore, the rift occupied by the ſlip muſt have been occaſioned by the rupture of the ſtrata, and afterwards filled up from above, by the ſubſtance of the ſlip: in the latter caſe, the uſual conſequence of the *ſlip* muſt have been counteracted by the inequality of the ſoil. In both the caſes obſerved, the ſubſtance of the ſlip was a grey indurated clay, with ſome ſuperficial traces of lead ore and ſulphur pyrites, and ſome calcareous ſpar. It is in vain, therefore, that volcaniſts, or rather plutoniſts, aſcribe theſe ſlips, and the diſorders that accompany them, to ſubterraneous eruptions.

eruptions. Muſt a mere indurated clay be deemed a ſubterraneoùs eruption? Can a maſs of it a few inches thick occaſion a great diſturbance, while one much more conſiderable, either occaſions none at all, or a much ſmaller? Would not a conic elevation of both the ſides of a diſrupted ſtratum be the natural reſult of ſuch an impreſſion from below? A poſition, however, which the diſrupted carbonaceous ſtrata never preſent.

The *courſe* or *direction* of the ſtrata has been obſerved to be moſt generally towards ſome point between E. and W. or N.E. and S.W. becauſe the winds from theſe points are moſt frequent, in Europe at leaſt; and by theſe and the viciſſitudes of heat and cold, moiſture and dryneſs, that ſeverally accompany them, diſintegration and decompoſition are moſt promoted.

The varieties of the *dip* ariſe from the inequalities of the baſis to which the ſtrata always conform. Morand, 62. When the dip forms what is called a horſe-ſhoe, deſcending from one mountain or hill, and aſcending on the oppoſite, it is becauſe the diſintegrated particles deſcended from both, and the waters found the ſame level. Laſtly, coal is ſcarcely ever found between calcareous ſtrata, becauſe carbon ſeldom exiſts in limeſtones, except that which forms a conſtituent part of fixed air; and calcareous mountains are leſs ſubject to diſintegration than thoſe of granite, porphyry, and trap. The origin of coal mines from diſintegration and decompoſition, which I have hitherto explained, is by far the moſt general, but in ſome particular caſes, coal ſeems alſo to have orginated, and ſtill to originate from tranſudation through ſubſtances *imperfectly decompoſed* or *recompoſed* after diſintegration. Of the firſt we have an inſtance in the argillaceous mountain of St. Georges near Milhant. 11 An. Chy. 272. at $\frac{3}{4}$ of the height of this mountain there is a ſtratum of coal which cuts the mountain in two; this muſt have been formed by tranſudation from the upper part before it had hardened; ſo alſo in the valley, between the mountains Juſſon and Chaminelle, in Dauphiné, there are ſeveral ſtrata of ſarcilite, compacted by a calcareous cement, and between them ſome inconſiderable ſeams of coal are found, which

5 ſeem

ſeem to have flowed from the contiguous calcareous mountains, for the ſtony maſſes of which theſe mountains conſiſt, preſent at intervals black ſpots, which ſmell like ſwineſtone. 11 An. Chy. 271. Moſt of the coal found in or under conſiderable maſſes of limeſtone, ſeems to have been formed in this manner. Near Kratigen, in Swiſſerland, a thick bed of coal is found in the midſt of limeſtone; but in the vicinity, petrol and ſoft bitumens are found in the limeſtone. Ferber Briefe Min. Inhalt. 24. We have, according to Mr. Genneté, a ſtill more evident inſtance of this mode of production of coal, from a ſubſtance ſubſequently recompoſed in the coal mines of Liege; there a ſpecies of ſandſtone is found, called by the country miners *agaz*, in which there are veins through which a bitumen, or rather a bitumen impregnated with carbon, tranſudes, which forms ſeams of coal at preſent, and in 40 years fill up the parts already worked. The maſs of this ſandſtone in the territory of Liege is, he ſays, to that of coal as 25 to 1; nay, Mr. Buffon, from whom I extract this paſſage, aſſures us, he ſaw himſelf ſome coal which had

 newly

newly tranfuded in this manner. 2 Buff. Mineral. in 8vo. p. 204, 206. In all thefe cafes it is plain, the fand originally formed by difintegration, had again by the operation of petrefcent juices, coalefced into a ftony fubftance, before the carbonaceous ingredient had been fevered from it. It alfo not unfrequently happens that the proportion of petrol is not fufficient to carry off the difengaged carbon, and in thefe cafes the coal muft remain difperfed in the cavities of the fandftone; hence the fhreds of coal often obferved in fandftone, but of a bad kind, not being fufficiently impregnated with petrol. Charp. 7, and 46. Hence alfo in many parts of Swifferland there is fcarce a fandhill without nefts of coal. 4 Helv. Mag. 119.

The connexion obferved between trap, bafalt, and coal mines, feems to indicate that thefe ftony fubftances contain fome proportion of carbon, and orginally a much greater. That hornblende does contain it has been already mentioned, as alfo that trap contains petrol. Bafalt, if it contains any carbon, contains leaft, but as trap abounds in hornblende is eafily decompofible, poſible, and generally accompanies baſalt, it is to its decompoſition that I think the ſtrata of coal found under it may be aſcribed; ſome have thought that the coal next under it, is always of that kind that difficultly burns, which the vulcaniſts aſcribe to its having been charred by the melted lava (ſo they call trap and baſalt) that flowed over it; but they are evidently deceived, for the coal next to the trap is frequently the moſt inflammable and bituminous, as at Meiſſen. 1 Berg. Jour. 1792, 379. The coal found under trap, is chiefly of vegetable origin, but above it true mineral coal which originated from the trap itſelf frequently occurs, as at Meiſſen, &c. This vegetable coal is formed of trees proſtrated by the deluge, accidently covered with trap, at its laſt period, as already ſhewn, and bituminated by the petrol flowing from the trap; hence at Meiſſen, its loweſt ſtrata, to which a ſufficient quantity of petrol could not reach, is the worſt. 1 Berg. Jour. 1792, 378. Carbonated wood is often found in various coal mines, not becauſe the coal derived its origin from wood, but becauſe the carbonic part of the wood was preſerved, and as it were embalmed by the petrol of the original coal mine, whereas in other places the wood decayed as usual.

To confirm this theory, I shall now proceed to shew its conformity to actual appearances, by proving the identity of the materials that form the strata, with those of the mountains, from whose disintegration and decomposition I have asserted the strata to have originated.

1°. The coal mines in the valley of Plauen, are skirted with mountains of argillaceous porphyry, and sienite, which frequently alternate with each other. Charp. 51. the sienite consists of hornblende and flesh red felspar, and very little quartz; the porphyry also contains red and bluish felspar, quartz minutely divided, and indurated clay, blue, red, and greenish. Charp. 50. 2 Berg. Jour. 1792, 153. accordingly the coal strata consist entirely of clay and sand of the same variety of colours. Some bivalved muscles, as well as vegetable impressions are found in some of these strata; the muscles are probably fluviatile, as the impressions are of fresh water plants; but I should not be surprised if marine shells should

ſhould alſo be found among them, as the porphyritic and ſienitic hills are ſurmounted with limeſtones, in which theſe ſhells are plentifully found. Charp. Ibid. and ſome of theſe might undoubtedly crumble down with the diſintegrated ſtones on which they repoſed; the wood probably grew on the mountains at the time of their decompoſition.

2°. The hills about Potchapel coal mines are entirely of argillaceous porphyry, containing only minute grains of quartz, and of which the felſpar is frequently converted into clay, hence the coal ſtrata here conſiſt entirely of clay. 6 Lempe, 41.

3°. From porphyraceous mountains, not very diſtant, the ſtrata of the coal mines in the valley of Planitz, placed between them, and of thoſe of Zwickau, Bockwa, and Reinſdorf, are evidently derived; the porphyry is argillaceous, and abounds in quartz, the quartz often contains mica, and is then called there a ſandſtone, its colour is often red, green, blue, or white, alternates with layers of clay, and red, ſoft, ferruginous, ſandſtone, and ſoon withers when expoſed to the air. Charp. 294.

Accordingly in the coal mines under the mould, we firſt meet grey and reddiſh clay, then a ſtratum of quartzy ſand 13 feet, then a fine grained ſandſtone, with organic impreſſions, and balls of agate or calcedony; under this, indurated clay, and ſhale, paſſing into the bituminous ſort; under this the coal lies. Ibid. 300, 301.

4°. The coal mines of Wilkiſchen near Kladrau, in Bohemia, are bordered by, and ſituated on the declivity of mountains of granite, and micaceous argillite; in the fields where they are found, blocks of granite repoſe on the ſurface. The ſtrata that cover the ſeams of coal conſiſt of a greyiſh white clay, micaceous ſand and ſhale, with vegetable impreſſions, which ſtrata, as Ferber juſtly obſerves, were formed by the diſintegration and decompoſition of the adjacent mountains. *Bohemia*, 300, 303*. 2 Berg. Jour. 1791, 56. all the ſtrata finally repoſe on granite, as he obſerved: the depth of theſe mines is only 36 feet.

5°. The chain of mountains adjacent to

* See the Engliſh tranſlation at the end of Born's Letters from Hungary.

Rive de Gier consist of granite and gneiss; the valley betwixt these and the river Gier, presents sandstones and breccias proceeding from their disintegration, 11 Ann. Chy. 270. and accordingly the coal mines are formed of strata of sandstone, but chiefly of shale, more or less bituminated, as we have already observed.

6°. The chain of mountains in Languedoc, that extends from Anduse to Villefort, consist of primitive sandstone, and abounds in iron mines; it is accompanied by a soil that originated from disintegration, and in this, coal covered with bituminous shale is found in heaps between the arenilitic rocks. Mem. Par. 1747, 1028. 1033. Mr. Tonson found the coal mine of Oedinburg in Hungary, covered with sandstone, but the neighbouring rocks were shistose mica. *Hungary*, 40.

7°. On the side of a granitic mountain, near St. Hypolite, Lavoisier found the following strata, red clay, black sandy earth, gravel mixed with felspar, gravel indurated into true granite, the same in a loose form, both these alternating with each other; shale, bituminous shale, with vegetable impressions,

preſſions, with thin veins of coal, and at laſt, granite. Mem. Par. 1778, 440. Here we evidently ſee the origin of coal from the mere detritus, or diſintegration of granite. Elſewhere alſo coal covered with ſandſtone repoſes immediately on granite, 2 Berg. Jour. 1790, 322. the ſandſtone then moſt probably proceeded from the granite. One of the mountainous ridges of Languedoc, called Servas, conſiſts of limeſtone richly impregnated with Petrol; now in one of the ravines of this mountain, beds of ſand, and of a highly bituminated coal, alternate with each other. Mem. Par. 1746, 1081, in 8vo. Is it not then certain that this mountain was degraded by diſintegration? That the bituminated coal ſeparated, and formed a diſtinct ſtratum, leaving the ſtony part in the form of ſand? And, that from ſucceſſive diſintegrations the multiplied ſtrata aroſe? Some ſhells are found over the coal, partly marine (viz. turbinites) and partly thoſe of land ſnails, Ibid. 1082. the former from the limeſtone, the latter muſt have been produced long after the mountain had emerged from the ſea: an exact confirmation of the preſent theory.

8°. Near

8° Near the river Angara, in Siberia, there are different ſtrata of coal, 3 or 4 inches thick, ſeparated from each other by argillaceous and arenaceous ſtrata, but without any organic remains. 38 Roz. 226. This river is ſurrounded by primitive mountains, from whoſe diſintegration theſe ſtrata evidently proceed; a diſintegration that probably preceded the production of organic ſubſtances.

9°. The dip of the ſeams of coal at Whitehaven is from E. to W. and of thoſe of Newcaſtle from W. to E. which ſhews that both proceeded from the interior intermediate mountains.

The practical inferences from this theory are,

1°. That coal is never to be expected in primeval mountains, as granite, gneiſs, &c. but that on the ſides of theſe, particularly if very high, or in the hanging level that ſlopes from them to ſome river or valley, it may be ſought.

2°. That there is ſtill a greater probability of finding it in the neighbourhood of mountains of argillaceous porphyry, as thoſe are ſtill more ſubject to diſintegration.

3°. That

3°. That it may be ſought with probability of ſucceſs in ſandſtone mountains, if ſandſtone and clay alternate, or ſandſtone, clay, and argillaceous iron ore.

4°. That in any elevated land in which ſandſtone and ſhale, with vegetable impreſſions, or indurated clay and ſhale, or bituminous ſhale, form diſtinct ſtrata, or clay, iron ore, and ſhale, with or without ſtrata of ſand, coal may well be expected.

5°. That if ſandſtone be found under limeſtone, or if they alternate with each other, and, particularly, if indurated clay and ſhale form any of the ſtrata, they afford a probable indication of coal; otherwiſe coal is very rarely found in, or under, limeſtone.

6°. That coal is very ſeldom found with argillite, and ſuch as has been is of the uninflammable kind.

7°. That where trap, or whin and clay, alternate, and more eſpecially trap and ſandſtone, coal may be expected; it is often, but not regularly, found under baſalt:—Wood coal is ſometimes found under both.

Laſtly, that coal frequently burſts out on the ſurface, or on the ſides of hills, in a withered

withered ſtate, which diffuſes itſelf to a diſtance from its origin, and requires an experienced miner to trace it truly to the ſeam to which it belongs.

ESSAY VIII.

OF COMMON SALT AND ITS MINES.

Common ſalt is found either in a liquid or in a ſolid form, and in greater plenty than any other ſalt whatſoever.

In the liquid form it is found in the ſea, in ſalt lakes, and in ſalt ſprings; as both theſe laſt, however, proceed from maſſes of ſalt contained in the earth, in a ſolid form, I ſhall poſtpone their conſideration to that of ſalt in a ſolid form.

§ 1.

Of the Sea.

The ſea, comprehending under that name the different oceans, the Mediterranean, Euxine, and Baltic, is undoubtedly the moſt conſiderable mine of any ſort exiſting in any known part of the globe, ſince judging from indiſcriminate experiments $\frac{1}{31}$ of its whole weight nearly, is ſaline matter;

ter; the extremes are $\frac{1}{32}$ and $\frac{1}{24}$, or 3 per cent. or 4 per cent. nearly. Phipps found the weight of its ſaline contents at the back of Yarmouth ſands nearly $\frac{1}{32}$, Lat. 53°. and Bergman found that taken up in the latitude of the Canaries to contain about $\frac{1}{24}$ of its weight of ſaline matter; but theſe quantities, even in the ſame latitude, are variable, there being leſs in rainy than in dry, and greater in ſtormy than in calm weather (if the degrees of heat be not very different), as ſtorms powerfully promote evaporation. Near land alſo, and particularly near the mouth of great rivers, it is evident the proportion of ſaline matter muſt be ſmaller than at a conſiderable diſtance from them. It has been generally imagined that at great depths the ſea water is more ſalt, though leſs nauſeous, than at the ſurface, but Mr. Bladh has completely proved, that to the depth at leaſt of 50 fathom, there is no difference. 35 Mem. Stock. Neither does the mere difference of latitude cauſe any conſiderable difference in the proportion of ſaline matter contained in ſea water, as appears by the experiments of Phipps, Baumé, and

Pages; and hence it is evident that the ſea does not derive its ſaltneſs from any mountains of ſalt contained in it, as ſome have thought, for in that caſe more would be daily diſſolved in the hot than in the cold climates. Phipps in latitude 80° N. 60 fathom under ice, found the water to contain of ſaline matter $\frac{1}{28,2}$, that is, 3,54 per cent. and Pages in latitude 81°, found 4 per cent. or $\frac{1}{25}$, the temperature I do not mention, as it could make no difference where ſuch minute quantities are concerned. In lat. 74° $\frac{1}{28}$ nearly, or 3,6 per cent. Phipps; 4,75 per cent. by Pages; but I fancy this muſt be a miſtake. In lat. 60° Phipps found $\frac{1}{25}$, and Pages $\frac{1}{28,5}$, or 3,5 per cent. in lat. 59° in the German Sea; in lat. 53° Phipps found $\frac{1}{32}$ on the back of Yarmouth ſands, but I ſuſpect ſome fault in this experiment, as the ſpecific gravity aſcribed to the water that contained it, does not at all agree with ſo ſmall a proportion of ſaline matter.

Lavoiſier ſcarcely got 2 per cent. from ſea water taken up, as he was informed, 4 leagues from Dieppe, and Monnet ſtill leſs. See Mem. Par. 1772, p. 349. in 8vo. and Monnet Nouvell. Hydrolog. 209. Either the

the water was not taken up at that diſtance from the ſhore, or they evaporated it too much, and too ſtrongly, by which means the marine acid, and even the ſalt itſelf, was, in great meaſure, volatilized.

Pages tells us, that from ſea water taken up in latitude 45° and 39 N. in the ocean, he obtained 4 per cent. of ſaline matter; and Beaumé analyſed ſome taken up by Pages in lat. 34° and 14°, and found each to contain about 4 per cent. and in lat. 25° N. Pages found ſea water to contain 3,75 per cent. and in latitude 10° 3,66; yet in lat. 4 it contained but 3,5 per cent. which agrees with Bladh's obſervations which I ſhall preſently mention.

In the ſouthern latitudes Pages found the proportion of ſaline matter ſomewhat greater.

Lat.....49°	50′ he found the ſea water to contain	4,1666 per cent. of ſaline matter.
.......46°	00′...............	4,5
.......40°	30′...............	4
.......25°	54′...............	4
.......20°	00′...............	3,9
But in l. 1°	16′...............	3,5 *

* See the Table prefixed to Vol. III. of Pages' Voy.

In the Mediterranean, the proportion of ſaline matter is ſaid to be much greater, namely, $\frac{1}{16}$, a quantity almoſt incredible; Dr. Hales is ſaid to have found it $\frac{1}{27}$, this proportion, however, ſeems too ſmall if the water was taken at a diſtance from the mouth of the great rivers *.

According to 2 Tournefort Voy. 410, the Euxine and Caſpian ſeas are leſs ſalt than the ocean, and ſo alſo is the Baltic, as we ſhall preſently ſee.

The ſpecific gravity of ſea water taken up at different latitudes, has alſo been examined with great accuracy by Mr. Bladh, of whoſe obſervations I ſhall here give an extract; he reduced his experiments to the ſpecific gravity the water would have at 20° of the Swed. Thermometer, that is, 68° of Fahr. and I have reduced them to that it would have at 62° of Fahr. ſuppoſing the dilatability of the ſalt water to be the ſame as that of diſtilled water, and in effect, Lambert found that even a ſaturate ſolution of common ſalt expanded between the temperatures of ice and boiling

* 1 Monro Min. Waters, 105.

water

water only $\frac{7}{1000}$ more than pure water, and the marine ſolutions being very weak muſt expand much leſs. The formula I followed is laid down in the firſt Volume of my Mineralogy. Bladh counts his longitude from Teneriffe.

Lat.		Long.		Sp. Gr. at 68°	Sp. Gr. at 62°
N.		E.			
59°	39′	8°	48′	1,0266	1,0272
57°	18′	18°	48′	1,0263	1,0269
		W.			
57°	1′	1°	22′	1,02669	1,0272
54°	00′	4°	45′	1,0265	1,0271
44°	32′	2°	04′	1,0270	1.0276
		E.			
44°	07′	1°	00′	1,02705	1,0276
40°	41′	0°	30.	1,0270	1,0276
34°	40′	1°	18′	1,0274	1,0280
29°	50′	0°	00′	1,0275	1,0281
		W.			
24°	00′	2°	32′	1,0278	1,0284
18°	28′	3°	24′	1,0275	1,0281
16°	36′	3°	37′	1,0271	1,0277
14°	56′	3°	46′	1,0269	1,0275
10°	30′	3°	49′	1,0266	1,0272
5°	50′	3°	28′	1,0268	1,0274
2°	20′	3°	26′	1.0265	1,0271
1°	25′	3°	30′	1,0267	1,0273
S.					
0°	16′	3°	40′	1,0271	1,0277
5°	10	6°	00′	1,0271	1.0277
10°	00′	6°	05′	1,0279	1,0285
14°	40′	7°	00′	1,0278	1,0284
20°	06′	5°	30′	1,0279	1,0285
25°	45′	2°	22′	1,0275	1,0281
		E.			
30°	25′	7°	12′	1,0273	1,0279
37°	37′	68°	13′	1,0270	1,0276

As the higher ſpecific gravity denotes the greater proportion of ſaline matter, hence he juſtly infers that in the Atlantic and Ethiopic oceans, the ſea is charged with moſt ſalt near the tropics, and becomes leſs impregnated near the line from the quantity of rain that uſually prevails there: we alſo ſee, that the ſouthern ocean is more ſalt than the northern, the reaſon of which probably is, that it is the original ocean in which all ſalt was at firſt contained, and by communication with which, the northern ſeas were ſalted, as ſhewn in Eſſay II.

By Wilke's experiments * it appears, that the Baltic is much leſs ſalt than the ocean, and that it is ſalter under a weſterly than under an eaſterly wind; and ſtill ſalter when a N.W. wind prevails. His experiments were made at 9° of the Swediſh thermometer, or 48° Fahr. I have reduced them to the temperature of 62°.

At 48°	1,0047	At 62°	1,0039	Wind at E.
......	1,0075	......	1,0067	Ditto at W.
......	1,0126	......	1,0118	Storm at W.
......	1,0105	......	1,0098	Wind at N.W.

* 33 Schwed. Abhand. 67.

Hence

ſome of the pits at Liege the ſtrata are, 1. mould, 2. clay, 3. a ſandy martial clay, 4. a calciferous ſand, 5. chalk with flints, from 42 to 72 feet thick, under this various beds of ſandſtone and ſhale. *Morand* Arts and Metiers, &c.

In the coal pits of Pirken, in Bavaria, ten ſeams of coal very near each other, alternate with ſhale, marl, and ſwineſtone; the ſwineſtone abounds in marine ſhells of all ſorts, and the marl ſtill more; theſe ſtrata form an angle with the horizon, of from 74 to 76 degrees. Flurl *Bavaria*, 103, 104.

CHAP. II.

FEW problems occur in the natural hiſtory of minerals of more difficult ſolution than that of the origin and formation of mineral coal and its mines muſt have been, before Lavoiſier had diſcovered that it was a conſtituent part of fixed air; but this degree of knowledge being gained, we muſt, of courſe, allow that carbonic ſubſtance is of equal antiquity with fixed air, and this air being a component part of many

Hence we clearly ſee, that the ſaltneſs of the Baltic proceeds from the weſtern ocean, and alſo perceive the influence of ſtorms.

The correſpondence betwixt the ſpecific gravity of the ſea water and its proportion of ſaline matter cannot be made out with much preciſion, as it contains two or three ſpecies of ſalts whoſe proportion to each other is variable, and Dr. Watſon's table, the beſt ſtandard for determining the proportion of ſalt, was formed only on ſolutions of common ſalt; yet as common ſalt is by far the moſt copious ingredient in ſea water, and as the Doctor owns that his ſalt was not perfectly pure, it will be found to determine the proportion of ſaline matter very nearly in all the caſes to which it reaches. It was formed at temperatures from 46° to 55°, hence I have ſuppoſed it at the medium of 50°, and calculated what the proportions relative to the preſent object ſhould be at the temperature of 62° in the following table.

Salt	Specific Gravity
Salt $\frac{1}{24}$	1,0283
... $\frac{1}{25}$	1,0275
... $\frac{1}{26}$	1,0270
... $\frac{1}{27}$	1,0267
... $\frac{1}{28}$	1,0250
... $\frac{1}{30}$	1,0233
... $\frac{1}{35}$	1,0185
... $\frac{1}{144}$	1,0033
... $\frac{1}{36}$	1,0105
... $\frac{1}{108}$	1,004
... $\frac{1}{261}$	1,0023

What degree of confidence we may repoſe in the indications concluded from Dr. Watſon's experiments, may be diſcerned in the following inſtances.

Briſſon diſſolved 2 oz. of the pureſt common ſalt in 16 oz. of diſtilled water, and found the ſpecific gravity of the ſolution 1,0790 in the temperature of 14° of Reaumur, that is, 63°,5 of Fahr. this ſolution then contained $\frac{1}{9}$ of its weight of ſalt; now by Doctor Watſon's table, a ſolution containing $\frac{1}{9}$ of ſalt has its ſpecific gravity 1,074, and this being reduced to that which it would have at 62° is 1,0732; here then the difference is $\frac{6}{1000}$, which I attribute to the ſuperior gravity of pure ſalt. Again, Bergman found the ſpecific gravity of ſea water

water taken up in lat. 28° N. near the Canaries, 1,0289 in the temperature of 15° on the Swedifh fcale, equal 60° of Fahr. and he found the faline contents $\frac{1}{27}$, and this fpecific gravity being reduced to the temperature of 62°, is 1,0286, and by the laft table we fee that Doctor Watfon gives it 1,0283 at 62° of Fahr. as the common falt was here mixed with other falts, we fee the agreement is nearer.

It is true Lord Mulgrave found the fpecific gravity of fea water taken up at the back of Yarmouth fands, to be 1,0280 in the temperature of 53° Fahr. or 1,0273 at the temperature of 62°, and yet he found the faline contents only $\frac{1}{31}$ nearly, whereas by Doctor Watfon's table, either the fpecific gravity fhould be 1,023, or if this were rightly taken, which I believe it to have been, the faline contents fhould exceed $\frac{1}{27}$. This difference I attribute to a fault in the conduct of the experiment, the evaporation having been either too quick, or too far pufhed, or both together; either way much of the falt, and of the acid of the earthy falts, muft have been loft. As to the fpecies of falts found in fea water, they

 may

may be reduced according to the moſt exact experiments hitherto made, only to three, viz. common ſalt, muriated magneſia, and ſelenite, now and then a very ſmall proportion of Epſom ſalt, and of aerated lime has been diſcovered. Bergman in the ſea water he examined and above mentioned, found 3,29 per cent. of common ſalt, 0,869 per cent. of muriated magneſia, and about ,001 of ſelenite, and the quantity of ſea water he examined was about 5,52 Engliſh pints, wine meaſure. Monnet in ſea water, taken up near Nantz, found 3,4 per cent. of common ſalt, nearly 1 per cent. of muriated magneſia, and about 0,0008 of ſelenite. In three other experiments made on ſea water, taken up near Dunkirk, Dieppe, and Granville, he found the ſame ſalts, and alſo a minute proportion of Epſom. Baumé *, who examined the ſea waters taken up by Pages in lat. 34° and 14°, found each to contain 4 per cent. of common ſalt, 0,12 per cent. of ſelenite, and a ſmall proportion of a ſalt, which he calls *magneſia of common ſalt*, and yet pretends

* Mem. Par. 1787, 547.

that its basis burns to *lime*. This salt he did not weigh, but decomposed it by an alkali, and found its basis dried at 212°, to amount to about 0,4 per cent. but the quantity of neutral salt formed by its acid he does not notice. He also found 0,12 per cent. of selenite, and asserts, that the sea water contained also Glauber, but in so minute a proportion *that he could discover none*, not recollecting that his calcareous salt would decompose it if there had been any; it is possible, however, that such minute proportions might coexist without meeting each other. Neither Lavoisier's nor Monnet's experiments on the sea water near Dieppe, are much to be depended upon, as neither of them appear to have operated on genuine sea water, for, as already said, that water scarcely contained, or at least they extracted from it, scarcely 2 per cent. of saline matter, of which far the greater part was common salt, and the next most copious part was muriated magnesia, and next to that selenite. Both found an exceeding minute proportion of Epsom, and Lavoisier says he found 0,08 per cent. of Glauber; but Monnet, though he operated

rated on the ſame quantity of water, viz. 20 pints, or 40 pounds, found no Glauber. On examining theſe analyſes, it is evident, that none of them, except Bergman's, were properly conducted.

I omit the experiments made by perſons unacquainted with the nature of magneſia, and am convinced that the experiments of all thoſe who attribute leſs than 3,8 per cent. of ſaline matter, or $\frac{1}{25}$, to genuine ſea water taken up in the northern ocean, unleſs in particular circumſtances, are faulty: the mean quantity ſeems to be 4 per cent. or $\frac{1}{25}$, the extremes $\frac{1}{24}$ and $\frac{1}{26}$; the ſpecific gravity of ſea water containing $\frac{1}{25}$ is 1,0275 in the temperature of 62°; Briſſon, it is true, found the ſpecific gravity of ſea water taken up in the bay of St. Brieux in Brittanny, to be only 1,0263 in the temperature of 63°,5 Fahr. which would be 1,0264 at the temperature of 62°. But as two ſmall rivers empty themſelves in that bay, and the water might have been taken up during the ebb, I look on this water to contain a mixture of freſh. Glauber's ſalt is ſaid to have been found in the water of the Mediterranean. Mem. Par. 1763, p.

326,

326, in 8vo. The Epſom, extracted from the ſea water at Harwich, Newcaſtle, Lymington, &c. is as to its acid part, artificial; the mother liquor of pyrites being added after the ſeparation of the common ſalt. That the ſea was originally much more ſalt than at preſent, will be ſhewn in the ſequel.

What was, or is, the primary cauſe of the ſaltneſs of the ſea, has been the ſubject of much diſcuſſion, and yet not only that cauſe, but even the reaſon why the particular ſalts already mentioned, and no other, exiſt in it, and alſo that, of the proportion to each other in which they are found, ſeem to me not difficult to explain, but in order to do ſo we muſt recur to the principles laid down in Eſſay I.

1°. That all ſubſtances incapable of farther analyſis, muſt be deemed ſimple until future experiments teach us otherwiſe.

2°. All ſimple ſubſtances muſt have been coeval with the creation, and have exiſted in the chaotic fluid, and, originally at leaſt, in an *uncombined* ſtate, the component parts of water alone excepted.

Now both the acid, and alſo natron or ſoda,

ſoda, the baſis of common ſalt, are ſimple ſubſtances in this ſenſe, as we know of no proceſs, either of art or of nature, by which either of them can either be decompoſed or formed. That ſoda is not the product of vegetation appears by the experiments of Mr. Du Hamel. Mem. Par. 1767, as he found the quantity of ſoda in kali to decreaſe progreſſively in ſucceſſive annual growths of the plant ſown in lands diſtant from the ſea. Both the acid and baſis muſt therefore have originally exiſted, though in an uncombined ſtate, in the chaotic fluid. Again, of all mineral acids, the moſt abundant are the marine and vitriolic; and of all baſes to which they may unite, calcareous earth, natron, and magneſia, are thoſe which are found in the greateſt plenty: iron, it is true, is found perhaps as plentifully as any of them, but then its combinations with either of thoſe acids would immediately be decompoſed by any of the above-mentioned ſubſtances. Farther, it cannot be pretended that the vitriolic acid is an original ſubſtance; it is evidently formed of ſulphur and pure air, and this air owed its extrication to volca-

nic

nic heat, as ſhewn in Eſſay I. therefore it could not have obtained as early a poſſeſſion of the above-mentioned baſes as the marine. This point is important, and may be proved in another manner: If the vitriolic and marine acids were coetaneous, they muſt have obtained a ſimultaneous poſſeſſion of the baſes to which they both have the ſtrongeſt affinity, viz. natron and calcareous earth, and, as after poſſeſſing the natron, there was ſtill a ſufficiency of calcareous earth to ſaturate the reſidue of both of them, there ſhould neither at firſt nor at preſent be any ſuch thing in the ſea as muriated magneſia, of which, however, it contains, as we have ſeen, nearly 1 per cent. but the ſea would at firſt contain Glauber, common ſalt, and muriated and vitriolated lime, and ſoon after (as Glauber's ſalt and muriated lime decompoſe each other,) it would contain common ſalt and gypſum, and at laſt, as gypſum is decompoſed by aerated magneſia, we ſhould have in the ſea at this day, common ſalt and Epſom ſalt only, or at moſt only common ſalt, Epſom and ſelenite, it being poſſible that there might not be enough of magneſia to decompoſe

compose all the selenite at first formed; but still we should find no muriated magnesia, and Epsom salt should be much more abundant than we find it, as it is still more soluble than common salt. Since, therefore, the sea does not contain these salts in the proportions just mentioned, but does exhibit a salt in a large proportion, which in the hypothesis of the coetaneity of the marine and vitriolic acids it could not contain, that hypothesis is plainly false *. On the contrary, if the marine acid be supposed to be more ancient than the vitriolic, (a supposition the truth of which has been proved,) then we shall find the saline contents of the sea to be just such, and in the same proportion, as experiments prove them to exist in it at present; for if marine acid was more ancient than the vitriolic, then it possessed the natron and calcareous earth previously to the existence of the vitriolic; and as muriated lime is decomposed by aerated magnesia, we should

* Perhaps the soda originally contained in the chaotic fluid helped the solution of siliceous substances, or, rather, maintained it for some time.

have

have common ſalt and muriated magneſia, and alſo muriated lime, as the aerated magneſia cannot be ſuppoſed to have been ſufficiently abundant to precipitate all the muriated lime after the exiſtence of the vitriolic acid: as this acid was at firſt ſulphureous, it could not decompoſe the common ſalt; but as there was a ſufficiency, and even a ſuperabundance of lime, or aerated lime, it is to this it would moſt eaſily unite, or if any of it were ſufficiently oxygenated to decompoſe the common ſalt, the vitriolated natron and Glauber's ſalt, thus formed, would immediately be decompoſed by the muriated lime, and thus the ſea would at laſt, after the interval of ſome years, poſſeſs what it now exhibits, common ſalt, muriated magneſia, and ſelenite; the common ſalt in the greateſt proportion, as it is the baſis to which the marine acid has the greateſt affinity; the muriated magneſia ſhould be the next moſt abundant, as it was formed by the decompoſition of the muriated lime; calcareous earth being the ſubſtance to which the marine acid unites moſt willingly after alkalies, and, conſequently, originally muriated lime exiſted in great abundance*: besides, muriated magnesia is exceeding soluble: Selenite would, indeed, if we consider the quantity of it originally formed, be more abundant were it not for its sparing solubility, for 500 or 600 parts of water are required to dissolve one part of it, consequently it cannot be as abundant in sea water as the more soluble salts. Hence also we see why no traces of alum were ever discovered in the sea, though it is much more soluble than selenite, as acids unite to argil less willingly than to calcareous earth or magnesia, and, therefore, none was originally formed in the chaotic fluid.

§ 2.

Of Common Salt in a solid Form, or Rock Salt.

Rock salt is found in immense masses in each of the four great divisions of the globe. In Europe it is found in England, Spain, Austria, Stiria, Hungary, Transylvania, Poland, Swisserland, &c.

The principal observations concerning it are,

* Hence, perhaps, the superior size of the most ancient shell fish.

1°. That

1°. That it is found moſtly in ſtrata, but ſometimes, though rarely, in veins, as near the banks of the Ebro, Bowles, 374. and at Friedenberg in the Dutchy of Salſburgh. 3 Mem. Lauſ. 388. and ſometimes it conſtitutes whole mountains.

2°. The ſtrata are parallel to each other, horizontal or undulating, commonly interlined with thin ſtrata of clay, or gypſum; thus they are found in Hungary, Born *Hungary*, 140. 144. And in Auſtria and Tranſylvania. 1 Gerh. 144. In Permia and Siberia. Maquart 67. In Swiſſerland. 2 Mem. Lauſ. and Wild's Tracts, and in ſome parts of Spain. Bowles, 376. The loweſt ſtrata are the thickeſt, the upper gradually thinner. Mem. Par. 1762, 1055, in 8vo. But ſometimes theſe heterogeneities, inſtead of lying between the ſtrata, are diſperſed in the ſalt, as at Wilickſa, and in England. Maquart, 59.

3°. That it is frequently mixed with bituminous clay, and ſometimes with that and pyrites, as in the county of Salſburgh, Bavaria, Stiria, &c. 3 Mem. Lauſ. and 11 Ann. Chy. 66.

4°. That it generally repoſes on a bed of

 indurated

indurated clay. Born. 3 Decouv. Ruſſes, 134.

5°. That it is almoſt always accompanied with gypſum, either *mixed* with the ſtrata that cover the ſalt, or *forming* one of thoſe ſtrata, or at leaſt conſtituting hills or hillocks in the vicinity of ſalt mines; this has been obſerved in Upper Auſtria, Hungary, and Tranſylvania, by Born. *Hungary*, 144. or mixed with it. Ibid. 165. And in Spain by Bowles, 164. By Count Razomuſki in Salſburgh. 3 Mem. Lauſ. 388, &c. By Guettard in the mine of Wilickſa in Poland. Mem. Par. 1762, p. 1055, in 8vo. By Pallas in that of Illetzki in Siberia. 3 Deſcouvertes Ruſſes, p. 419.

6°. Gypſum is alſo conſtantly found under the indurated clay on which ſalt rock repoſes. 2 Mem. Lauſ. Part II. 7.

7°. It is generally found among ſecondary ſtrata, either in valleys, plains, or mountains, on the deſcent, or at the foot of, or ſurrounded by, primary mountains of granite, gneiſs, primitive limeſtone, &c. At Arzew near Algiers, the ſalt pits are ſurrounded by mountains, in winter they are a lake, in ſummer ſolid. Shaw's Travels,

229. The ſalt mines of Wilickſa are in an elevated valley at the foot of the Carpathian mountains, and thoſe of Salſburgh, Bavaria, Calabria, and Spain, are at the foot of mountains. Mem. Par. 1762, p. 1069, in 8vo.

But it alſo exiſts in primeval mountains in Siberia, as ſalt lakes are found in them, per Pallas, 1 Born, Phy. Arbeit, 21, and in the deſert primitive plain of Cobea. 1 Act. Petrop. 38. formed at the earlieſt period of the exiſtence of the globe, when the ſea was abundantly more ſalt than at preſent.

8'. It is alſo found, though very rarely, at great heights, though never at the ſummit of a hill: it is ſaid that the ſalt mine of Arbonne in Savoy, is nearly in the ſnowy region; more commonly at conſiderable depths. The mine of Torda in Hungary, lies at the depth of 36 feet, Born, 141. that of Wilickſa at the depth of 500. Mem. Par. 1762, 8vo. p. 1055. that of Durenberg at the depth of 1320 feet. 3 Mem. Lauſ. 290. Some others in Siberia are within a few feet of the ſurface. At Norwich at the depth of 120 feet. 3 Jars. 332. In ſome parts only 100. 2 Phil. Tranſ. abr. 523.

9°. The ſtrata that cover rock ſalt are

mould, marl, clay more or leſs indurated, frequently coloured and bituminous, ſand often micaceous, limeſtone, gypſum, ſandſtone with an argillaceous cement.' At Peſackna in Hungary, the ſtrata are mould, black ſpotted clay mixed with mica, micaceous ſand, black ſtrong ſmelling clay, under which the ſalt repoſes. 1 Gerh. Geſch. xiii. So alſo the ſoil that ſurrounds and covers the mine of Illetſki in Siberia is ſandy. 1 Herman, 38, 39, &c. 3 Deſcouv. Ruſſes, 134. Near Mingranilla in the province of Valentia in Spain, there are hillocks of ſalt covered with gypſum. Bowles from ſeveral adjacent circumſtances concludes, with great probability, that the gypſum was originally covered with limeſtone to the height of 800 feet, which was waſhed off by ancient inundations. The whole is on a deſcent, for you deſcend to arrive at the hillocks. *Spain*, 164. 166. At Wilickſa the ſtrata are ſand, clay or marl ſlate, limeſtone: even when limeſtone does not form one of the ſtrata, it is at leaſt found in the neighbourhood, or ſurrounding the ſalt mines. At Norwich the ſtrata that cover the ſalt are ſand and indurated (perhaps bituminated) clay or ſhale. 3 Jars. 332.

At

At Torda, in Hungary, the upper ſtratum is alſo indurated clay; at Mingranilla, gypſum.

10°. Marine remains have frequently been detected in the ſtrata that cover rock ſalt; thus at Wilickſa, ſea ſhells are found in the clay that forms one of the ſtrata that cover it. Mem. Par. 1762, p. 1055, in 8vo. and the bones of land animals. 8 Lempe Mag. 47. nay madrepores and bivalves are found in the ſalt. Maquart 51. 16 Roz. 463. the concomitant limeſtone alſo contains marine petrifactions.

11°. Many mountains entirely conſiſting of ſalt have been diſcovered. The ſalt mountain of Cordona in Valentia, is from 4 to 500 feet high, and about three miles in circumference. Bowles 406. Fortis mentions ſeveral in Calabria, attended with ſome of gypſum, ſeveral in the ſtates of Algiers and Tunis are mentioned by Shaw, p. 229. and another in the province of Aſtrachan. 3 Buff. Min. 8vo. p. 371. the ſalt in this, however, contains a mixture of foreign ingredients, the nature of which has not been accurately determined. The ſalt of the mountain Jibbel Hadiffa is of a pur-

 pliſh

pliſh colour and bitter, but whether the bitterneſs proceeds from glauber, or muriated lime, or magneſia, or ſome two of them, is not known, but that it proceeds from one or other of them is certain, as this bitterneſs is eaſily waſhed out. In the province of Yakoutz, in Siberia, near the river Kaptindei there is a mountain of ſalt 180 feet high, and 120 in length, but at $\frac{2}{3}$ of its height it is covered with a ſtratum of red clay, which reaches to its ſummit. 1 Gmelin Voy. 342, cited by Maquart, 82.

Patrin ſuſpects that many granitic mountains contain ſalt, which, he thinks, has been the cauſe of the deſtruction of many of them, and at this day promotes the decompoſition of many that ſtill exiſt; hence he derives the ſalineferous, ſandy plains of Siberia. 4 Nev. Nord. Betyr, 167, 174. but it more commonly, at leaſt, proceeds from ſalt ſprings beneath the ſand. See 1 Herman *Uber die Uraliſch. Erze Gebirge*, 36.

12°. Rock ſalt is of various colours; white, red, orange, purple, blue, and green; the *white* contains argil, and muriated magneſia, rarely gypſum and muriated lime; the

the *red* and *orange* contain gypſum and glauber; the *blue* mangancſe, and the *green* copper, per Haſſenfraz, 11 An. Chy. 74. theſe colours are, however, ſometimes optical illuſions.

13°. The quantity or maſs of ſalt already diſcovered in many mines is enormous. The mine of Torda is from 30 to 40 fathom thick, its form, where worked, circular, and its diameter 14 fathoms; that of Coloſer has its thickneſs 60 fathoms, and its diameter 50. Born. *Hungary*, 140, 143. to ſay nothing of the parts not yet worked. The mine of Karaulnaia Gora, in Siberia, is 60 fathom long, 9 or 10 broad, its depth, as yet unknown, it having been worked only to the depth of 3 feet. 3 Deſcouv. Ruſſes, 145. The famous mine of Wilickſa, in Poland, is, according to Mr. Coxe, 6691 feet long, 1115 broad, and 743 deep, as far as its extent is known. 1 Coxe, 197. The ſtratum of rock ſalt at Norwich, in Cheſhire, is 50 feet thick. Dundonald on the Salt Manufacture, p. 2. Add to this, the ſalt mountains I have juſt mentioned, and the vaſt quantities of ſalt yearly extracted in different parts of the world from ſalt

 ſprings,

ſprings, and ſalt lakes, almoſt all of which flow from maſſes of rock ſalt.

On weighing the various circumſtances juſt mentioned, it muſt appear very evident that rock ſalt derives its origin from the ſea, and that the ſpaces which it now occupies were originally vaſt hollows ſucceſſively filled with ſea water at diſtant intervals during the period of the diminution of the level of the ſea, to nearly its preſent height, that is, antecedently to the general deluge. Its diviſion into ſtrata; the paralleliſm of theſe ſtrata; their horizontality, where the level of the baſis would allow of it; the undulations obſerved on its ſurface; the marine ſhells found, not only in the ſtrata that cover it, but even between thoſe of the ſalt itſelf; the thin beds of argil, and particularly thoſe of gypſum, intercepted very frequently between its ſtrata; all beſpeak a marine origin, and its inland ſituation, and the great heights at which it is ſometimes found, prove its depoſition to have happened during the retreat of the ſea.

To form a juſt idea of the mode of its formation, we muſt obſerve, that the ſea originally

originally contained much more faline matter than at prefent, for it is now deprived of thofe vaft faline maffes that exift in every quarter of the globe, as fhewn in the 12th obfervation, which originally had exifted in the chaotic fluid; it is therefore not a harfh conclufion, that it originally contained twice more faline matter, than it at prefent does, and fince at prefent, it contains about $\frac{1}{20}$ of its weight at a medium, that originally it contained $\frac{1}{15}$ part of its weight of falt. This being premifed, we may fuppofe the hollows that form its prefent mines, to have been originally filled by the fea, and afterwards abandoned for *fome* time after its retreat. In this interval the aqueous part would naturally evaporate, and form a bed of faline matter, but at the return of a *fpring tide* thefe hollows would again be filled; the clayey particles would fubfide before this fecond portion of water could be evaporated; and the gypfum would alfo be depofited long before the cryftallization of the falt; and thus the beds of clay and gypfum intercepted between the ftrata, according to the fecond obfervation, would be formed. Hence it would neceffarily follow,

follow, that the lowest stratum of salt should be the thickest, for in proportion as the cavity was filled with deposited salt, in the same proportion its capacity for containing sea water was diminished, and consequently the succeeding strata were progressively thinner. It may be asked, why all hollows and cavities did not equally become salt mines, since the sea water during its retreat must have equally filled all of them? The answer is easy: those cavities could only become salt mines, whose bottom and sides contained sufficiently dense strata of clay, particularly bituminated clay, these alone could detain the sea water, and prevent its percolation; hence as strata of argil are very rare in primitive mountains, the hollows of primitive mountains, except in a few rare instances, contain none. Hence also gypseous mountains, hills, or hillocks, are frequently found without salt; the salt is never found without gypsum, unless the gypsum were washed off by subsequent inundations, for as gypsum requires a large portion of water for its solution, it would be deposited long before the water could either evaporate or run off, and therefore

therefore did not require that the ſea ſhould be reſtrained by clay, or bituminated clay; both the time and the evaporation requiſite to its depoſition is 100 times ſmaller than requiſite for the depoſition of common ſalt. The various ſtrata that at preſent cover the mines of rock ſalt, proceed from the diſintegration and decompoſition of the primitive mountains, at whoſe feet they lie, after the retreat of the ſea; hence, not only ſea ſhells, but the bones of land animals, are ſometimes found in them, according to the 10th obſervation. With reſpect to the mountains of ſalt mentioned in the 11th obſervation, there is great reaſon to think that they were alſo originally formed in vaſt cavities, but that the earthy and ſtony ſubſtances that formed theſe cavities were carried off by ſubſequent inundations; it deſerves to be remarked, that theſe mountains have been obſerved only in countries bordering on the Mediterranean and Caſpian, namely in the province of Valentia, in Spain; near Algiers, in Africa; in Calabria; aud in the province of Aſtrachan. Now it is well known, that both theſe ſeas had once ravaged all the neighbouring coun-

tries

tries with irresistible violence, and probably diminished, though not destroyed, the saline mountains. Hence no hillock of gypsum, or any trace of it, is found near the salt mountain of Cordona, and Bowles expressly tells us that the salt mine of Mingranilla was once covered with earth and stone to the height of 800 feet, though at present it is only covered with a slight stratum of gypsum, all the superior earthy and stony matter having been swept away by floods. *Spain* 164, 166. The great and cautious geologist Saussure observed in various instances, indications of the destruction of mountains by inundations. 2 Sauss. 127. 5 Sauss. 441. 6 Sauss. 91, 154, 244. that near the Kaptindei, in Siberia had evidently its sides torn off by an ancient inundation. Some may perhaps imagine, that if salt mines were formed before the universal deluge, their salt should be dissolved in that immense mass of waters; to these I would observe, that the salt mountain of Cordona has been exposed to rain ever since the deluge, and even washed by a river, and yet it still subsists; the other saline mountains above mentioned must also have withstood

innumerable

innumerable floods of rain, ſuch as generally fall in hot countries, for ſalt, though when pulveriſed and agitated it be ſoluble in three times its weight of water, and in a ſhort time, yet when in large maſſes and at reſt, requires a very conſiderable ſpace of time to diſſolve it, even in ten times its weight of water; thus Bergman found ſalt kept in agitation, ſoluble in 10 times its weight of water in one minute, but when at reſt the ſame quantity of ſalt (only 4 drams) required 34 hours for its ſolution. 5 Berg. 114. In Tranſylvania a rivulet of freſh water flows over a ſtratum of ſalt without any contamination of ſalt, being preſerved from it by a ſtratum of earth which it depoſits. Wild, *ſur les Salines de Bex*, 120. See alſo 2 Deſcouv. Ruſſes, 34. In many inſtances, bitumen flowing from the neighbouring mountains, muſt have entered into theſe cavities when filled with ſea water, and during its evaporation; this mixing with the clay already contained in the water, or that accompanied the bitumen, muſt have enveloped the ſalt, and prevented its regular cryſtallization, and after the bitumen had flowed out,

out, the inſpiſſated ſaline maſs muſt, by ſuperincumbent preſſure, have been forced into the veins through which the bitumen had flowed, and thus the ſalt veins in the mountains of Salſburgh, Bavaria, &c. may have originated.

Among the many fanciful hypotheſes of Mr. Buffon to explain the origin of minerals, no inſtance occurs in which the powers of his imagination have proved ſo evidently inſufficient to impoſe even upon himſelf, as in his attempt to explain the origin of ſea ſalt. This we muſt infer from the glaring contradictions into which he falls.

In the firſt volume of the ſupplement to his Natural Hiſtory *, he tells us, that acids and alkalies ſhould rather be conſidered as *products of art*, than as natural ſubſtances, but in the third volume of his Mineralogy (a ſubſequent work) he tells us, the ſea was originally acidulous, and acquired its alkali only from the deſtruction of *organized* ſubſtances †. Before we advance further, let us examine this poſition; theſe organiſed ſub-

* Page 66, 8vo edition. † Page 347.

ſtances muſt be either of the animal or vegetable tribe. Now no marine animals could live in the ſea before it became ſalt no more than they now can in freſh water, and ſtill leſs in acidulous water; it was not therefore from any alkali reſulting from the deſtruction of theſe, that it became ſalt, nor, indeed, are fiſh known to contain any; neither could it derive its ſalt from the deſtruction of vegetables, for the kali and other vegetables that contain the baſis of ſea ſalt, acquire it from the ſea, and contain none when at a diſtance from it; but his embarraſſment does not end here. According to him, 1 Epoques, p. 132. Anno Mundi, 30 or 35000, the globe was ſufficiently cooled to permit the condenſation of the watery vapours that ſurrounded it; the ſea was thus formed, it covered the earth to the height of 13000 feet, and nothing but the ſummit of ſome few mountains ſurpaſſed it. Yet at this period, he ſays, the ſea began to be peopled with ſhell and other fiſh, with which no ſpecies at preſent known exactly correſponds, as *cornua ammonis*, &c. of a prodigious ſize, and theſe ſubſiſted from anno mundi 30 or 35000 to

anno

anno mundi 40 or 45000, when the ſea, which at firſt was too hot for our preſent race of fiſh, became too cold for thoſe ancient ſpecies. Here he ſeems to have forgotten the origin of ſalt; for if theſe fiſh exiſted when the ſea covered the whole earth to the height of 13000 feet, where could the vegetables have grown from whoſe deſtruction the acidulous ſea, he ſays, acquired the baſis of common ſalt? He has even deprived himſelf of the reſource of deriving this baſis from kali, or any known vegetable, for he tells us, that the earth was at this period too hot to bear any known ſpecies of vegetables *. He even aſſerts that fiſh and vegetables originated *at the ſame time*. How then could the ſea have acquired the ſalt neceſſary for the exiſtence of fiſh? even if the Andes, Mont Blanc, the Tartarian mountains, and a few others riſing above 13000 feet, did bear kali or ſome other imaginary alkaliferous vegetables, How could the ſea acquire this alkali, ſince, according to him, it never covered them? and if it had, How many millions

* 1 Epoques, p. 140, 141.

of

of years would be requiſite for the production of ſo much as enters into the compoſition of all the ſalt now in the ſea, and in the earth ? The vaſt ſize of the moſt ancient ſpecies of fiſh he aſcribes to the great heat which he gratuitouſly ſuppoſes the ſea to have originally poſſeſſed, whereas it may, with greater probability, be attributed to the greater quantity of ſalt and calcareous matter, as ſelenite, originally contained in it ; in this reſpect, at leaſt, it is certain that the ancient ſea differed from the preſent. No other geologiſt has attempted to account for the origin of the alkaline part of ſea ſalt; none, except Mr. De Luc, has felt the importance of ſearching into the origin of the ſaline maſs at preſent contained in the ſea *; Dr. Halley imagined

* Il eſt indubitable pour tout geologue attentif et éclairé, que la mer actuel eſt le reſidue d'un liquide dans le quel ſe ſont formés tant nos ſubſtances minerales que notre atmoſphere, il n'eſt donc pas indifferent davoir preſent à l'eſprit que ce reſidue contient encore la magneſie, la terre calcaire, l'acide vitriolique, et probablement d'autres ſubſtances tenues que nous ne connoiſſons pas encore, et dont la connoiſſance pourr ait nous conduire à des deſcouvertes ſur les operations paſſées, &c. 40 Roz. 362.

that the ſaltneſs of the ſea proceeds from the quantities daily carried into it by rivers, and, conſequently, that it conſtantly increaſes, but if they anciently conveyed into it no more than they do at preſent, an innumerable ſeries of ages would have been requiſite to render it as ſalt as it now is. Surely no river can convey into it as much as the river Cardonero, which waſhes a mountain of ſalt, and yet at the diſtance of nine miles from the mountain not a particle of ſalt can be diſcovered in its water, and Bowles, who remarks this peculiarity, affirms he made a number of experiments on the water of rivers at their mouths, and never could diſcover any, though ſometimes he had elſewhere detected $\frac{1}{1000}$ part of ſalt. *Spain*, 407. 409. we muſt, therefore, ſuppoſe, as he was not ignorant of chymiſtry, that he uſed the great teſt of marine acid, the ſolution of nitrated ſilver: this diſcovers 1 grain of ſalt in 43000 grains of water, and if a river waſhing a mountain of ſalt does not carry off $\frac{1}{43000}$ part in ſo warm a climate as that of Spain, How little can other rivers contain not placed in ſuch favourable circumſtances? How few paſs through any ſtrata of the earth containing ſalt? How few, therefore, can we ſuppoſe to contain it in any proportion ever ſo minute? Dr. Rotheram found that Thames water, taken up at Billingſgate, impregnated, as it may be ſuppoſed to be, with ſea ſalt from the quantity uſed in London, and waſhed into it by various drains, does not contain $\frac{1}{1000}$ part of ſalt. 2 Watſon's Eſſays, p. 99, 100.

Of Salt Springs and Lakes.

Geologiſts, in general, have paid but little attention to ſalt ſprings and lakes; they have ſuppoſed their contents to be the ſame as thoſe of the ſea, though in fact they are very different either with reſpect to the ſpecies of ſalt they contain, or their proportion, or both; this will appear from the few inſtances I ſhall adduce out of many; for ſalt ſprings, in particular, are much more numerous than the known mines of rock ſalt: there are many in England, as in Cheſhire, Worceſterſhire, Staffordſhire, Hampſhire, &c. Several in France, Swiſſerland, Stiria, Siberia, &c. Germany alone is is ſaid to afford upwards of three hundred *. Hungary and Siberia abound in ſalt lakes.

The circumſtances accompanying ſalt ſprings are much the ſame as thoſe attendant on rock ſalt; theſe ſprings are found iſſuing chiefly from ſecondary ſtrata, and the hills they proceed from are frequently gypſeous, or at leaſt contain that ſubſtance; they are often ſurrounded by limeſtone ſtrata, and theſe generally cover indurated clay. Sometimes they iſſue from gypſum, as at Frankenberg. Charp. Saxony, 377. Sometimes from indurated clay; ſometimes from limeſtone or marl, rarely from ſandſtone. 2 Mem. Lauſ. Part II. 8. Deſcript. Pyrenées, 25. Sometimes they lie very deep, and are covered with various ſtrata; thus at Altkoſen, they lie at the depth of 575 feet; the uppermoſt ſtratum is clay 5 Saxon fathom thick; the next bluiſh grey ſtone-marl 24 fathom, under this is a browiſh red looſe marl 10 fathom, in which are ſome thin widely ſeparated ſtrata of gypſum; this is ſucceeded by 41

* 2 Mem. Lauſ. Part II. p. 5.

fathom

fathom of lamellar gypſum, with a few interrupted ſtrata of clay, and, laſtly, 2,5 fathom of pure lamellar gypſum, under which the ſalt water is found. Charp. 377. Sometimes they burſt forth from very elevated ſituations, as thoſe of Spain. Deſcript. Pyren. 25. At Shonebach, ſalt ſprings were diſcovered at the depth of 240 feet; at Dorrenberg, at the depth of 720. 1 Gerh. Geſch. 134. The ſaliniferous hill Konigſhorn in Weſtphalia, conſiſts of marly limeſtone repoſing on ſemiprotolite; in the adjacent plains there are ſeveral ſalt ſprings. 1 Klaproth, 356. The ſalt ſpring of Pyrmont iſſues from a reddiſh iron ſhot micaceous ſandſtone; it lies deep, and is raiſed by a pump. 1 Weſtr. Abhandl. 281. Some have ſaid that ſalt ſprings are never found above rock ſalt, but always lower. 2 Mem. Lauſ. Part II. p. 11. the contrary, however, has been obſerved in England, for ſalt ſprings were diſcovered both above and below the level of the Norwich bed of rock ſalt, the upper being impregnated by reſting on the bed of ſalt, and the lower by running from it. 2 Watſon, 39.

Moſt of the ſalt ſprings in Germany,

 Salſburgh,

Salſburgh, Stiria, and Swiſſerland, particularly the famous ſprings of Aigle in the Canton of Berne, are thought to proceed from ſalt diſperſed through, or involved in bituminous indurated clay or gypſum.

Salt ſprings are frequently diſcovered by the vegetables that grow near them, and which are peculiar to ſaline ſoils, as Triglochin Maritimum, Salicornia, Salſolakali, Aſtertripolium, Glaux Maritima, &c.

In Siberia ſalt ſprings have been found in valleys betwixt hills of jaſper, 3 Deſcouvertes Ruſſes, 161, and thoſe of Navarre iſſue from primitive limeſtone. Deſcript. Pyren. 25, 26. hence they are not ſtrangers to primitive mountains.

Klaproth analyſed ſeveral of the ſalt ſprings of Konigſhorn in Weſtphalia, he found them* all to contain, beſides common ſalt, from $\frac{1}{315}$ to $\frac{1}{725}$ of muriated lime, a ſmaller proportion of ſelenite, but only $\frac{1}{14500}$ of muriated magneſia, the proportion of common ſalt was only $\frac{1}{18}$ at moſt; though theſe ſprings flow through marl, yet as aerated lime does not decom-

* 1 Klaproth, 355, &c.

poſe

poſe muriated magneſia, the almoſt entire abſence of this ſalt cannot be derived from that cauſe; the rock ſalt therefore, from whence this ſpring flows, muſt have been depoſited before the muriated calx originally contained in the ſea had been decompoſed by aerated magneſia.

Weſtrumb analyſed the ſalt ſprings of Pyrmont, and found them to contain, beſides common ſalt, alſo muriated magneſia, and a very minute proportion of Glauber, and ſelenite; but in the old ſpring of Luneburgh, he found a much larger proportion of theſe ſalts; for 100 parts of this brine, by his analyſis, contained 25 of common ſalt, 0,2 nearly of muriated magneſia, and ſomewhat leſs of Glauber and ſelenite.

Hellot analyſed the ſalt ſprings of Montmorot in the year 1760, a period at which the nature of magneſia was not known in France; beſides common ſalt he found in them both Glauber, ſelenite, and muriated (magneſia) deliqueſcent ſalts, with ſome particles of bitumen. Mem. Par. 1762, 546, in 8vo.

The quantity of ſalt yearly extracted from ſalt ſprings is aſtoniſhing, as the few inſtances I ſhall now mention abundantly ſhew; a ſingle pit at Norwich yields at an average, 4000 ton, or 80000000 pounds. 2 Watſon, 41. the ſprings at Koningſhorn, in Weſtphalia, 11850511 pounds. 1 Klaproth, 359. the old ſalt ſpring of Luneburg yields 75600 gallons of brine in 24 hours, and computing the gallon to weigh 8lb, (and it muſt be more) it yields 604800 lb. of brine, of which ¼ is ſaline matter, almoſt entirely common ſalt, equal 151200lb. or above 55 million yearly. 1 Weſtr. Abhandl. 295. What then muſt be the yield of all the other known ſprings in different parts of the world, and how immenſe the quantity of common ſalt?

Of Salt Lakes.

Salt lakes are evidently the reſult of the accumulation of the ſtreams of ſalt ſprings; they are found principally in Siberia and the Crimea, ſome alſo in Ruſſia, and many in Africa, often on the ſummit of mountains. *

* As on Bogdo. 2 Deſcouv. Ruſſes, 30. and in the Crimea. 1 Bergb. 301, 302.

On

On the mountain of Tſchernayer, one of thoſe that form the chain of Inderſki, there is a ſalt lake environed by hillocks of gypſum, and its bottom is clay; the ſalt frequently cryſtallizes in the ſhallow parts into a ſolid cake by reaſon of the evaporation during the ſummer, for it is ſupplied by ſubterraneous ſalt ſprings. Pallas 3 Deſcouv. Ruſſes, 419, 430. that of Buſskunzatzkoi on the banks of the Volga is about 30 miles in circumference. 2 Deſcouv. Ruſſes, 34. Gmelin mentions another called Jamicha, whoſe water is ſaturated with ſalt. Voy. p. 100. all have gypſum in their vicinity. Maquart 84. 2 Deſcouv. Ruſſes, 31. Near Schaſkojam, in Ruſſia, there are three ſalt lakes ſupplied by ſalt ſprings. 1 Deſcouv. Ruſſes, 45, 46. In the neighbourhood of Aſtrachan there are ſeveral ſalt marſhes. 3 Deſcouv. Ruſſes, 85. the Caſpian itſelf may be regarded as a vaſt ſalt lake, receiving its ſalt from the neighbouring ſprings, 3 Deſcouv. Ruſſes, 80. it contains rather more of glauber than of common ſalt. Ibid, 85. The lake Aſphalt, otherwiſe called the dead ſea, contains more ſaline matter than any other known lake or water; Lavoiſier found it to contain 44 per cent, that is 6 of

common

common ſalt and 38 of a mixture of muriated magneſia and muriated lime; he found its ſpecific gravity 1,24061, but in what temperature he does not tell, probably at 10° of Ream, Mem. Par. 1778, p. 69, 71. Briſſon at the temperature of 63,5 of Fahr. found it 1,2403. *Gravités Specif*, 399.

No exact analyſis has been given of the waters of any other lake; but from the exiſtence of Glauber's ſalt, and muriated lime in ſeveral of them, it is plain that their compoſition is very different from that of the ocean. The Glauber found in the Mediterranean, and in the waters of Peccais that communicate with it*, is eaſily accounted for, it being the product of the numerous volcanoes that actually exiſt in its iſlands, or vicinity. 3 Bergm. 268. but in the neighbourhood of ſeveral ſalt ſprings and ſalt lakes the exiſtence of volcanos has not even been ſuſpected, its origin muſt, therefore, be deduced from other cauſes. With reſpect to the Caſpian, theſe are not difficultly traced; this ſea waſhes the feet of ſeveral mountains containing pyrites and

* Mem. Par. 1763 p. 326, in 8vo.

other

other ſulphurated ores*; theſe in a ſtate of oxygenation either fall or are waſhed into it, and thus, much of the common ſalt muſt in a courſe of ages have been decompoſed; the traces of ſuch accidents ſtill exiſt, as vitriols have been diſcovered buried in the ancient ſandy bed of that ſea†. In other caſes the Glauber might have ariſen from a mixture of native ſulphur, or pyrites and common ſalt; that pyrites frequently penetrate or ſhoot through various ſtones of all the genera, is well known; by long expoſure to the air and moiſture they are at laſt vitriolized, and their acid preys on any baſis to which it has an affinity. Thus Werner in an old mine at Scharfenberg found ſelenitic cryſtals in the rifts of a heap of granite dug out of a mine 200 years before, and in that part only of the granite which had withered by expoſure to the air, and, conſequently, the cryſtals were of recent formation: the matrix of the mine was a calcareous ſpar, and pyrites ſtill exiſt in it, hence the origin of the vitriolic acid is evident. 2 Biblioth. du Nord. 73. 77.

* 3 Deſcouv. Ruſſes, 83. † Ibid. 93.

Again,

Again, in a cavern near Hildesheim, one side of which is formed of red argillite, and the other of bituminous marlite, with which it is also vaulted, and the whole covered by limestone, Glauber is found crystallized in the argillite; as there are salt mines near Hildesheim, it is probable that some existed in the argillite; the marlite also frequently contains native sulphur, and thus the existence of the vitriolic acid cannot be ambiguous. per Hofmeister, 1 Chy. Ann. 1790, 46. The Glauber thus formed is itself frequently decomposed, and the natron which formed its basis set free, or at least left in a hepatic state; this decomposition arises from a mixture of petrol or bitumen, which, in a long course of time, gradually deoxygenates the vitriolic acid, and leaves it in the state of sulphur. Thus in the territory of Debrezin near the lake Bogod in Hungary, natron is found in a state of efflorescence, but mixed with a greasy substance, and some proportion of Glauber, Mem. Berl. 1770, 13. 16, 17, and in the neighbouring lakes of Derctske and Soboslo nothing but Glauber is found. 2 Chy. Ann. 1795, 126. In the neighbourhood

hood of the lakes, Glauber is found mixed with petrol, and frequently in the ſtate of a hepar, ſee Pazmand's Treatiſe in Vol. I. of Waſſerberg's Diſſertations, p. 417. On the coaſt of the ſea of Azof there are ſalt marſhes, but the ſalt is mixed with ſulphur, and hence the ſalt is often in an hepatic ſtate. 1 Bergbau, 301. From theſe hepars the ſulphur may gradually be chaſed by fixed air, and thus the alkali ſet free from it. Patrin alſo found liver of ſulphur in ſome ſalt lakes of Siberia. 4 Nev. Nord. Beytr. 196. In other places, however, the decompoſition of common ſalt cannot be attributed to vitriolic acid: thus natron is found in vaſt quantities in India, in pits by the ſea ſide, mixed with a conſiderable proportion of brown ferruginous earth. Tranſ. of the Society of Arts and Manufactures for 1788, Vol. VI. Mr. Keir examined ſome parts of this natron which had been refined, and found it to contain 58,8 per cent. of mild alkali, 17,2 of common ſalt, and only 24 of water, the unrefined contained 1 per cent. of the earth, which paſſes through the filter, impedes the cryſtallization, and alters the ſhape of the cryſtal, (in (in this respect it resembles the imperfect alkali mentioned by 3 Bergm. 268). Here, as no trace of the vitriolic acid is found, I attribute the decomposition of the common salt to the ferruginous clay, which in a hot climate may in a long course of time decompose it, for in a *strong* heat clay is known to possess this power, particularly if ferruginous, and what a *strong* heat may effect in a *short* time, a *weaker* heat may effect in a *long* time; hence, as Mr. Bergman observes, the natron found in the earth, both in India and Africa, is free from common salt at the surface, but becomes contaminated by it so much the more, as it descends deeper *.

3 Berg. 267.

ESSAY IX.

ON METALLIC MINES.

To treat this ſubject as amply as even the preſent ſtate of our knowledge concerning them would admit, is not my intention; ſuch a treatiſe would require an entire volume: my purpoſe is ſingly to give a general idea of their formation in the different ſtates in which they are known to exiſt; and of the ſort of mineral *ſoil*, if I may ſo call it, or country in which particular kinds of them are moſt generally ſituated.

Moſt metals (under which name I here comprehend alſo ſemimetals) are found in four ſtates, native, ſulphurated, calciform, or ſaliniform.

§ 1.

Of Native Metals.

The metallic ſubſtances moſt commonly found in a native ſtate, are gold, ſilver, platina,

tina, mercury, copper, biſmuth, nickel, and arſenic, more rarely iron, antimony, tin, and lead.

Originally, it is probable, all metals exiſted in the chaotic fluid perfect and uncontaminated, as alſo in a minuter ſtate of diviſion than any earthy ſubſtance, and all (except iron) in a much ſmaller quantity; this minuter ſtate of diviſion I infer from their greater diviſibility at this day, for a minuter quantity of a metallic ſubſtance diſſolved in any menſtruum may be detected by appropriated precipitants than of any earth: their high ſpecific gravity ſeems to proceed from a cloſer union, and, conſequently, a minuter diviſion of their integrant particles; their diſtance from each other, (ariſing from the ſmallneſs of their quantity,) prevented them from uniting with each other as readily as the earths; the minuteneſs of their diviſion kept them long ſuſpended, and with the earths none of them could combine for want of affinity, while in their metallic ſtate; this ſtate, however, after the extrication of the oxygen, could not long continue.

The cryſtallized ſtate in which they are often

often found, is alſo a full proof of their minute diviſion and of their ancient ſolubility in water; if it were by any acid their ſolution was effected, as ſome have imagined, that acid would ſtill be found in thoſe cryſtals. The various ramifications which they exhibit when diffuſed through ſtony ſubſtances proceed from the diſtribution of the rifts which they fill up.

Native Gold.

Of all metals gold is moſt frequently found native. According to Bergman, it is more univerſally diffuſed than any other metal, except iron; this may be a conſequence of its great diviſibility and want of affinity to other ſubſtances, as oxygen, ſulphur, &c. hence at the emerſion of primeval mountains, it remained entangled or diſperſed through the ſtony maſſes of many of them wherever theſe were permeable to water; the golden particles were, however, in a courſe of ages, waſhed and carried down in minute rills into the neighbouring plains, until arreſted by ſome obſtacle long enough to ſuffer the gold to depoſit;

these minute particles being thus brought into contact in the minutest state of division, united with each other by virtue of their integrant affinity, sometimes involving sandy particles, and thus formed those shapeless masses, of various sizes, which are sometimes met with in various countries, and lately in the county of Wicklow; that these lumps were never in fusion is evident from their low specific gravity, and the grains of sand found in the midst of them. I found the specific gravity of a lump found in the county of Wicklow, of the size of a nutmeg, to be only 12,800, whereas after fusion it was 18,700, and minute grains of sand appeared on its surface. Hence many rivers were anciently auriferous, which now cease to be so; as the Tagus, Po, Pactolus. Heber. Pliny lib. xxxiii, cap. 4. and though in France some are still auriferous, yet it appears, by the testimony of Diodorus, that they were much more abundantly so in former ages*. Hence also native gold is seldom alloyed with any metal, except silver or copper, to which it

* Lib. v. cap. 19.

has the greateſt affinity, and which are alſo leaſt liable to a combination with ſulphur or acids.

It is oftener found in iron ores than in any other, becauſe theſe are far more univerſal and abundant than any other, particularly in a more or leſs indurated and brittle, brown or reddiſh brown iron ſtone; though originally it was depoſited in primeval mountains, yet from them by ſubſequent operations of nature it has frequently been depoſited in ſecondary maſſes, yet ſtill it is moſt frequently found in quartz, felſpar, &c. ſometimes in gypſum, baroſelenite, &c *.

Native Silver.

Silver, in proportion to the quantity of it that exiſts, is much more ſeldom found native than gold, by reaſon of its affinity to ſulphur; it was depoſited in the ſame manner as gold, and its cryſtallization proves it was once in a diſſolved ſtate. The metals it is moſt commonly alloyed with, and to which it has the greateſt affinity, are cop-

* Werner *Gange*, 150.

 per,

per, and antimony, in the abſence of lead, mercury, tin, and biſmuth. The ſtony ſubſtances it affects moſt are quartz, baroſelenite, calcareous ſpar, fluors, ſiderocalcite, more rarely hornſtone, flints, agates, aſbeſtus, ſteatites, and phoſphorite. It is rarely and only in minute quantities in ſecondary mountains. Lehm. flotz. Geſch. 210.

Native Copper.

In proportion to the quantity of copper in other ſtates, native copper is ſtill rarer than native ſilver, though there is ſcarce any mine in which ſome quantity of it has not been detected; it is more frequent in Siberia than elſewhere. It is ſeldom abſolutely free from ſulphur. The ſtony ſubſtances it affects moſt are quartz, baroſelenite, ſometimes zeolyte, fluors, and gneiſs. It is frequently accompanied with red iron ochre, red copper ore and malachite. Its affinity to ſulphur, ſufficiently accounts for its ſcarcity. It is ſometimes found in ſecondary mountains. Lehm. flotz, 210.

Native

Native Iron.

This is ſtill ſcarcer than the foregoing, as it is eaſily oxygenated and ſulphurated. The vaſt maſſes found in Siberia and Peru, ſeem to have originally been agglutinated by petrol, and left bare, when the ſurrounding earthy or ſtony maſſes either withered or were waſhed off.

Native Biſmuth.

Generally found with ores of grey or bright white cobalt ores.

Native Arſenic (See poſt Arſenic Pyrites).

Found only in the veins of primeval mountains (Emmerling) Lehm. flotz, 126. ſo alſo Miſpickel. Lehm. Ibid. accompanied by red ſilver ore, or galena, orpiment, ſulphurated cobalt, ſulphurated nickel, ſparry iron ore, grey copper ore, ſulphur and copper pyrites.

Native Mercury.

In clay in ſecondary mountains. Lehm. flotz, 214.

§ 2.

Of Sulphurated Ores.

The formation of ſulphurated ores was formerly conſidered as ſcarcely explicable, ſulphur being ſuppoſed inſoluble in water; hence many imagined, that depoſited at Aix la Chapelle, and other hot ſprings, not to have exiſted in the waters, but to have been formed out of them; Bergman, however, removed this difficulty, and has ſhewn that hepatic air is really contained in ſuch ſprings; my own experiments have ſhewn that diſtilled water takes up $\frac{2}{3}$ of its bulk of this air, and conſequently that a cubic foot of water may well contain 1152 cubic inches of this air, which are equal to 374 grains of ſulphur; now water takes up ſulphur when in an aerial ſtate, for no other reaſon, ſurely, but becauſe when in that ſtate it is moſt minutely divided; therefore originally, when ſulphur was in the minuteſt ſtate of diviſion it was equally ſoluble in water, and ſtill continues to be ſo when in the ſame circumſtances. Hahneman found that 1000 grs. of cold water would take up 2,3 of ſulphur when this was in the ſtate of hepatic air. § 67 in note. Again, metallic ſubſtances are ſoluble in water various ways; thus calx of lead is ſoluble in alkali, and lime, and precipitable therefrom by hepatic air. 1 An. Chy. 53. and is more attracted by lime than by other earths. Ibid. and hence lead mines are often found in limeſtone; ſo alſo mercurial calces are ſoluble in lime water. Ibid, 61. Even the perfect metals when ſufficiently divided are ſoluble in water; thus Cronſtedt has ſhewn that ſilver is taken up by water, and depoſited on the ſurface of rifts in Chriſtiana Pit, in Norway. 17 Mem. Stock. 272. Mercury is alſo in ſome degree ſoluble in water, for water that has long boiled over it gains an anthelmintic power; ſo alſo is regulus of arſenic, for water takes up $\frac{1}{1100}$ part of it, as Hahneman has ſhewn. Uber die Arſen. Vergift. § 28. and the ſolubility or ſuſpendibility, (as ſome may chooſe to call it,) in mere water may be affirmed of all other ſubſtances when ſufficiently divided, for they differ only in degree; when therefore water impregnated with ſulphur meets

 with

with that charged with any metal, it precipitates with and unites to that (gold and platina excepted;) the ſulphurated metal thus precipitated in the ſtate of the minuteſt diviſion, is itſelf ſoluble in water, as may be proved in many inſtances. Baron Trebra found pieces of wood that had been left 200 years in an old mine covered with calcareous ſpar, black or vitreous ſilver ore and red ſilver ore, both of which are ſulphurated. 1 Chy. An. 1786, 77. Brinkman found copper pyrites newly formed in a heap of the refuſe of copper pyrites which had anciently been roaſted, and out of which the copper had been extracted. 1 An. Chy. 1785, 264. Three circumſtances have contributed to deceive thoſe mineralogiſts and geologiſts that have denied the ſolubility of ſulphur, metals and ſulphurated, ores; the firſt is that after boiling perhaps a few grains of thoſe ſubſtances in perhaps 1000 times their weight of water, they could not after long boiling, find any perceptible loſs of weight in them, not conſidering that their ſcales could not diſcover a loſs of $\frac{1}{1000}$ part of a grain, which perhaps was all that 1000 parts water could diſſolve.

diſſolve. The ſecond is, that having pulveriſed them as minutely as could be effected by art, they thought them ſufficiently ſo, whereas they ſhould perhaps be divided into parts equal to the millionth part of a grain. The third, that they did not conſider that recent precipitates are much more ſoluble than thoſe whoſe conſtituent or integrant parts have been united for a conſiderable time. Thus Hahneman found that common orpiment requires 5000 times its weight of water to diſſolve it; but that recently formed by precipitation with hepatic air, required only 1000 times its weight of water, though it contained a much larger proportion of ſulphur than common orpiment. Uber Vergiſt. § 34. Arſenico ſulphurated waters have been detected in Cornwall. 6 Phil. Tranſ. abr. Part II. p. 186, 187. and antimony has been found in a ſtate of ſolution in the mine of Santa Cruz de Mudela in Spain. Bowles, 26. Beauman alſo found the walls of an old mine at Munſtcrappel ſmeared with native cinnabar that had tranſuded. Berolding, 174. 179.

Mountains, as ſhewn in Eſſay I. were formed

formed by a more or leſs perfect cryſtallization; thoſe leaſt perfectly cryſtalliſed thus far reſembled thoſe formed by mere depoſition, that after a certain degree of deſiccation their maſſes were capable of a much cloſer approach to each other, or of what builders commonly call *ſettlement*. From ſuch ſettlement cracks and rifts muſt have enſued, many alſo may have proceeded from earthquakes, ſee Advers. N. 162. Theſe rifts originating from the formation of the ſtrata themſelves, and not from the relation or poſition of the ſtrata with reſpect to each other, muſt have pervaded and croſſed them in various directions, ſuch rifts are called *veins*. To theſe as to a lower point the waters ſtill remaining in the minute interſtices of the ſtony maſſes muſt have faced, ſlowly conveying all the particles which remained uncombined with the ſtony maſſes, as being ſuperfluous to their formation, or for want of affinity, and yet ſufficiently comminuted to be ſoluble in water. Of theſe, the moſt ſoluble were firſt carried off, namely, the *earthy*, and being depoſited on the ſurfaces of the rift, formed, what are called, the *veinſtones*.

Subſequent

Subſequent infiltrations gradually carried into them the leſs ſoluble particles, namely, the *metallic*; in ſome caſes, however, the metallic particles ſeem to have been firſt depoſited, probably becauſe the earthy were far more diſtant. With reſpect to ſulphurated metals in particular, whether they were conveyed in their ſulphurated ſtate, or whether theſe rifts were at firſt ſtored with hepatized water, by which the metallic ſolutions were gradually precipitated as ſoon as they arrived; or whether, on the contrary, the rifts were firſt filled with metallic ſolutions, and the metals gradually precipitated by the acceſs of hepatized water, I ſhall not pretend to decide; probably each caſe might have taken place according to the various circumſtances that might have occurred; ſulphurated ſtreams are even now found in the mines of Cornwall, 6 Phil. Tranſ. abr. Part II. 185, and fraught with earthy matter, and that ſulphur and vitriolic acid may be found in granite, ſee 2 Sauſſ. 300.

The cryſtallization obſervable in theſe veins both of ſtony and metallic, or ſulphurated metallic ſubſtances, was the conſequence in the one caſe of the diminution of the watery

menſtruum,

menſtruum, and in the other of ſlow precipitation. The waters were ſlowly diminiſhed, partly by evaporation, which Baron Trebra has ſhewn to take place in the interior of mountains *, and partly by ſlow dribbling through the minuter rifts of the bottom of the veins. Where ſuch rifts do not occur, the waters depoſited their contents between the ſtrata of which the mountains conſiſt. Hence we ſee why metallic veins ſeldom occur in granitic mountains, or thoſe of jaſper and the harder ſtones, as their texture is too cloſe to permit the percolation of water, at leaſt in ſufficient plenty, and becauſe their rifts were previouſly occupied and filled with ſtony maſſes, as being more ſoluble, and therefore ſooneſt conveyed into them; thus ſilex ſufficiently comminuted, is ſoluble in about 1000 times its weight of water, or even leſs †; whereas metallic ſubſtances require much more; but if the granitic ſtones are

* See his firſt Letter, Vom. Innern der Gebirge.

† Thus Stucke found that 20 oz. of water, or 9600 grains, contained 14 grains of ſilex. Chy. Unterſuch. 119. It was accompanied with aerated magneſia and argil.

in a ſtate of decompoſition, as in the lower mountains they often are in Cornwall, &c. there they may be metalliferous. On the other hand, gneiſs, and ſhiſtoſe mica, argillaceous porphyry, and argillites, being much ſofter, are the principal abodes of metallic ores.

In Saxony tin ore is found between the layers of granite. 1 Lempe Mag. 103. 106. For a farther account of *veins*, I muſt refer to Werner's and Gerhard's celebrated Treatiſes, as they are not the direct object of this Eſſay.

The formation of *entire ſtrata* of ſulphurated ores ſeems more difficult to explain on the principles here ſtated, as they occupy ſpaces not previouſly empty, but filled with another ſubſtance. This difficulty will, however, vaniſh when it is conſidered,

1°. That the matter which at firſt poſſeſſed the ſpace now occupied by the ſulphurated metallic ſubſtances, was, at the time of its excluſion, in a looſe or ſlightly coherent ſtate.

2°. That the ſulphurated ores are more or leſs perfectly cryſtallized.

3°. That cryſtallizing ſubſtances exert a force

force indefinitely ſtrong in removing all obſtacles. Thus water on the point of cryſtallizing into ice, burſts even metallic veſſels; and thus alum cryſtallizing by inſenſible evaporation, in the midſt of a turbid ſolution of clay, repels the clayey particles from the ſpace occupied by its cryſtals.

4°. That the ſulphurated ores thus found in ſtrata, are never ſo free from heterogeneous mixtures as thoſe found in veins *.

Sulphurated Silver Ores.

Theſe moſt commonly accompany each other; thus the *vitreous ſilver ore* is accompanied often by the red, alſo by galena, or blendes, or mixed with ſulphur pyrites, or iron ochre, or ſiderocalcite, or hepatic pyrites, ſee 2 Pabſt's Catal. and Leſke's. The *red* ſilver ore is often accompanied by the brittle antimoniated, or mixed with iron pyrites, or blende, or ſiderocalcite, or bright white cobaltic ore, or a ſmall portion of galena, or copper pyrites, the *light red* often with native arſenic or orpiment.

* Laſius, 423. Monnet Mineralog. 265, 266.

The

The ores generally mixed with each other are thoſe, 1°. That have the ſame mineralizer, and ſame degree of ſolubility. 2°. Thoſe which are ſoluble in different degrees, but which in point of local ſituation are neareſt each other. Theſe ores are always in veins, and never found forming ſtrata, and yet ſcarce ever in *granitic* mountains. Werner *Gange*, 165. Gerh. Mineralog. 222.

Sulphurated Copper Ores.

Copper pyrites is frequently accompanied with grey copper ore, or mixed with arſenical pyrites or galena, or black or brown blende, or ſulphur pyrites.

It is found both in veins and ſtrata. Wideman. But chiefly in veins. Gerh. Min. 231.

The *grey copper ore*, beſides antimony, which it always, and lead and ſilver, which it often, contains, is moſtly accompanied by copper pyrites, or galena, ſparry iron ore, ſulphur pyrites, brown and yellow blende. It is found in ſecondary, as alſo in primary, mountains, but chiefly in theſe laſt. 2 Emmerling, 244. per Lehman, that which

which contains silver is never found in secondary mountains. Lehm. Flotz. Gesch. 125.

Black copper ore is found with copper pyrites, grey copper ore, malachite or green calx of copper also with the purple copper ore, or iron ochre.

Sulphurated Iron.

It occurs in strata in bellies and in veins; the striated frequently accompanies galena; it is itself accompanied by siderocalcite, or the sparry iron ore, or ochre, or galena, rarely (as at Menidot near St. Lo in Normandy) by cinnabar. Monnet Vitrioliz, 15, 16. That in veins has been found accompanied with cobalt and silver ores. Charp. 211.

Sulphurated Tin.

Is accompanied by copper pyrites, it is very rare, and found only in primary or derivative mountains.

Sulphurated Lead.

It is frequently accompanied with brown, or yellow, or black blende, copper or sulphur

phur pyrites, arſenical pyrites, or ſparry iron ore; ſometimes with red ſilver ore, or calamine, cobalt or native arſenic. It is found in primary and ſecondary mountains, but oftener in primary. Gerh. Mineral. 247.

Sulphurated Zinc.

The blendes are generally accompanied with galena, grey copper ore, ſulphur pyrites; the black, often with the above, or arſenical pyrites, or copper pyrites; more rarely with the white iron ore, or magnetic iron ſtone. Found only in primary mountains, Lehm. Flotz. 126.

Sulphurated Mercury.

Accompanied with iron ſtone, ſulphur pyrites, and arſenical pyrites, ochre, native mercury, galena, green, blue, or grey, copper ore. Berolding, 209, &c. In Idria it forms ſtrata. Lehm. 220. In the Palatinate and Idria it is found in ſecondary mountains, but that of Almaden is found in veins in ſandſtone, which appears to be primitive. Bowles, 40, &c.

Sulphurated Antimony.

With ſulphur pyrites and iron ochre; chiefly, if not ſolely, found in primary mountains. Lehm. Flotz. 126. 220. There are very few mines of it known.

Sulphurated Cobalt.

With ſulphurated nickel, native ochre, red calx of cobalt, native biſmuth, ſulphurated ſilver ores, copper and arſenical pyrites, and ſulphur pyrites, vitreous copper ore, galena, blende, ſparry iron ore. Found both in primary and ſecondary mountains, particularly in the *troubles* of theſe laſt. Lehm. 220.

Sulphurated Biſmuth.

With native biſmuth, ſparry iron ore. tinſtone, arſenical or copper pyrites.

Sulphurated Nickel.

With ſulphurated cobalt, and ſulphurated ſilver ores.

Arſenical Pyrites.

Found only in primeval mountains, oftener in layers than in veins. Emmerl. Lehm. 126.

126. Accompanied by tinſtone, galena, black blende, ſparry iron ore, ſulphur and copper pyrites.

Orpiment.

Found principally in ſecondary mountains; yet, per Lehm. Flotz. 126, only in primary; but he contradicts this in p. 214. Is not accompanied by any other ore but realgar.

Realgar.

Principally in primary mountains, accompanied by native arſenic, galena, red ſilver ore, ſometimes by cobalt, ſulphur pyrites, grey copper ore, brown blende, grey and red manganeſe.

Sulphurated Uranite.

Accompanied by galena, copper pyrites, iron ochre, ſometimes by bright cobalt ore, red calx of cobalt, or vitreous ſilver ore.

Of Calciform Ores.

Calciform ores are formed by infiltration. 10 Buff. 197.

Calciform Copper Ores.

Accompanied by each other, alſo by copper pyrites, vitreous copper ore, and brown iron ſtone, and iron ochre.

Calciform Iron Ores.

Magnetic iron ſtone frequently forms layers in primitive mountains, as in gneiſs and ſhiſtoſe mica, nay, it forms entire mountains; ſometimes it is alſo found in ſecondary mountains.—Accompanied by ſulphur, or magnetic pyrites, or copper or arſenical pyrites.

Specular iron ore found partly in veins, partly in layers, and, perhaps always, only in primitive mountains. Emerling. Accompanied by compact red iron ſtone, magnetic iron ſtone, ſulphur pyrites, rarely with copper pyrites, or arſenical pyrites, or galena, or tinſtone.—*Compact red iron ſtone* found in veins and layers accompanied by red hematites.

Red hematites found in veins, but principally forming conſiderable layers.

Compact brown iron ſtone found in veins, and vaſt ſtrata in ſecondary mountains. Voight

Voight Pract. 113. Accompanied with ſparry iron ore, or brown hematites, rarely with ſulphur pyrites, or copper ores.

Sparry iron ore found in primary mountains, *but only in veins, or veinſtones*; alſo in ſecondary mountains. Accompanied by brown iron ſtone, or ſiderocalcite, and very rarely ſilver ore. Deſcrip. Pyren. 13.

.Upland *argillaceous iron ſtone* found in ſecondary mountains only. Accompanied by calamine, or galena, ſometimes with ſulphur pyrites.

Emeril, in Spain it is found in ferruginous ſandſtone, or in iron ſtone.

Calciform Lead Ores.

The white accompanied by each other, and galena, iron ochre, ſulphur and copper pyrites, calciform copper ores, ſometimes by brown iron ſtone, or brown blende.

Calciform Tin Ores.

Found in veins, or layers, only in primeval mountains, as granite, gneiſs, ſhiſtoſe mica, and ſcattered in the diſintegrated maſſes of theſe mountains, but never in

 calcareous.

calcareous. Werner *Gange*, 165. Lehm. 125.

Accompanied with ſulphur, copper, or arſenical pyrites, wolfrom, molybdena, iron ochre, black, and rarely with brown blende. Seldom or never with ſilver ores, lead ores, or cobalt ores, or with baroſelenite, calcareous ſpar, or gypſum.

Native Turpeth.

Found with galena in Morsfeld. Berolding, 219.

Calciform Zinc Ores.

Found only in ſome ſecondary mountains, Emerling, never in granitic, gneiſſy, or ſhiſto micaceous. Werner Gange, 165. Accompanied with iron ochre, galena, white lead ore, compact brown iron ſtone.

Calciform Cobalt.

Found principally in ſecondary mountains, but alſo in primary. Accompanied by other cobalt ores, iron ochre, native ſilver, many copper ores, ſulphur pyrites, the *red* accompanying many other ores.

Calciform Manganese.

The grey, or bluiſh, or browniſh black, is frequently accompanied with red or brown compact iron ſtone, or ſparry iron ore.

The red is found with the gold ores of Nagaya, brown blende, galena, grey copper ore, copper pyrites, realgar.

Tungſtenite.

Found only in primitive mountains.

Uranitic Ores.

Micaceous, accompanied by compact red and brown iron ſtone, iron ochre, ſometimes with olive copper ore, or black or yellow cobalt ores.

Some veins were originally open, as appears from the rounded ſtones and petrifactions found in them; thus in the granitic mountain of Pangel in Sileſia, there is a vein whoſe inclination is 70°. filled with globular baſalt. 2 Berl. Beob. 197. So alſo the veins of wacken, called *butzen wacken* in Joachimſtahl in Bohemia, in which trees and their branches, &c. are found.

 1 Chy.

1 Chy. Ann. 1789, 131. See alſo Werner *Gange*, § 44, &c. But that all veins were originally open to day and filled from above, ſeems to me improbable. 1°. Becauſe ſulphurated ores could not thus be formed, as the waters containing them would immediately be precipitated by contact with the atmoſphere, and nothing but a ſlow, gradual, and ſucceſſive, precipitation, ſkreened from the atmoſphere, can account for their cryſtallizing in perfect cryſtals of conſiderable ſize. 2°. Becauſe inundations conveying only metallic particles, without any mud yet flowing from the ſurface of the earth, is highly improbable, and yet they muſt have been ſuch, elſe all veins would be filled with depoſited mud. 3°. Regulus of arſenic, which is found in many veins, could not thus be conveyed, as it would be immediately oxygenated by expoſure to the atmoſphere. 4°. Becauſe the quantity of ſulphurated metals that could be conveyed in a ſtate of ſolution, as it muſt have been to cryſtallize if held in pure water, would be too ſmall, even in the greateſt veins, to afford any conſiderable cryſtallization, ſo that numberleſs ſucceſſive inundations one

knows

knows not whence, and successive evaporations must be supposed. 5°. In some veins the ore is found adhering chiefly to the roof, as in that of Philip Ludwig at Gablau in Silesia. 1 Gerh. Gesch. 251. Must not that have been formed by percolation? 6°. Because we are led to such supposition by no proof or analogy, whereas the theory of successive percolation is grounded on the observation of metallic substances being thus conveyed at this day, as already shewn. Some mines are found at great heights, as that of Crumhubell in Silesia, which contain galena and silver ore. Mem. Berl. 1771. Whence could the inundation proceed that could overflow and fill them? And how often must it have been renewed for a series of years before any considerable quantity of metal could have been deposited?

The objections to the theory of percolation do not seem to me of any weight: The first objection, namely, that veinstones of different species cross each other in the same mountain, only shews that particles of stone of different species were dispersed in different parts of the *same* mountain, and were aſſembled by water flowing in different paſſages and at different Æras. The ſecond objection, that veinſtones, if firſt formed, would obſtruct the paſſage of metallic particles into the vein, is juſt the ſame that Charpentier and Gerhard made to the theory of petrifactions; for the calcedonic particles, for inſtance, that form the interior of many ſhells muſt have traverſed the calcareous, 6 Sauſſ. 83. and in ſome inſtances even ſiliceous; yet this objection is, in this inſtance, juſtly diſregarded. Beſides, how many veinſtones are ſtill penetrated with metallic particles? Are not cryſtals of calcareous ſpar found in the midſt of balls of agate? Laſius, 264. See 1 Berl. Beob. 372. and in quartz, 7 Sauſſ. 84. The third objection, namely, that waters could not circulate below the bed of rivers, is equally, and obviouſly, inconcluſive: numberleſs particles of water that penetrate mountains may have no connexion with the bed of rivers. How many ſtalactitic concretions are found in the caverns of Kilkenny, Derbyſhire, &c. far below the level of the adjacent rivers? The fourth objection, that no metallic particles are found ſound in the rocks adjacent to veins, may be anſwered by ſaying, that the metals have been long ago carried out of them; beſides the proportion may be ſo ſmall as to elude all reſearch, as in the mines of Ramelſberg. See Schlutter, Chap. IX. hence gold is not found in the mountains, out of which, nevertheleſs, it is yearly conveyed by inundations, and 2 Sauſſ. 411. Mem. Par. and ſilver as in Cronſtedt's obſervation. Moreover all the advantages ſo ingeniouſly deduced by Werner, from obſerving the various antiquity of metallic veins, and of their contents, are equally applicable to the theory of percolation.

Veinſtones, originating only from the minute particles detached from any combination, and having (except lime) no particular connexion with any metal, are indifferently found with them.

ORES FOUND IN PRIMEVAL MOUNTAINS.

The ores found in primeval mountains, are principally and commonly the following:

In

In Granitic Mountains.

From 2d Lenz. Flurl Bavaria, 256.

Tinſtone, galena, compact brown iron ſtone, ſulphur pyrites, hematites, blende, biſmuthic, cobaltic, arſenical pyrites, and molybdena, rarely magnetic iron ſtone, 1 Chy. Ann. 1797, 111. hence copper and ſilver ores ſeem excluded, though native gold, ſilver, or copper, are not.

In Gneiſs.

Silver ores, martial, copper and arſenical pyrites, magnetic iron ſtone, lead ores, tin ores, blende, and cobalt.

In Shiſtoſe Mica.

Ores of iron, copper, tin, lead, cobalt, and antimony.

In Stellſtein.

Ores of copper, or ſilver, or lead.

In Hornblende Slate.

Sulphur pyrites, copper ore. Aikin *Wales*, 119.

In Argillites.

Ore of ſilver, copper, lead, ſulphur pyrites, blende, calamine. Aikin Wales, 21, antimonial and mercurial ores; hence ores of cobalt, &c. are excluded. The great belly of ore in the Paris mountain is found under aluminous ſlate. Aikin *Wales*, 136.

In Shiſtoſe Chlorite.

Sulphur pyrites, octohedral magnetic iron ſtone.

In Grunſtein.

That formed of hornblende and mica ſometimes affords ores of copper, or iron, or ſulphur pyrites.

In Clay Porphyry. (Saxum Metalliferum Bornii.)

Ores of ſilver, copper, iron, lead, blende, and antimony.

In Hornſtone Porphyry.

Sparry iron ore, ſulphur pyrites, galena, black blende, biſmuthic ores.

In

In Serpentine.

Specular iron ore, magnetic iron ſtone, ſulphur pyrites.

In Trap.

Specular iron ore, magnetic iron ſtone, ſulphur pyrites.

In Wacken.

Iron ſtone, ſulphur pyrites, mangneſe.

In primitive Limeſtone.

Magnetic iron ſtone, ſulphur pyrites, copper pyrites, galena, yellow blende.

No veins are found in ſandſtone, or hornſlate. Charp. 27.

IN SECONDARY MOUNTAINS.

In compact Limeſtone.

Copper pyrites, ſparry iron ore, ſulphur pyrites, galena, cinnabar. Bowles, 64.

Marlite.

Cinnabar. Berolding, 45.

In

In bituminous Marlite. (Lehm. 218.)

Copper ores.

In Chalk.

Pyrites. Monnet Mineralog. 339.

In ſiliciferous Sandſtone.

Cinnabar.

In argillaceous Sandſtone.

Vitreous copper ore, malachite, iron ſtone, ſulphur pyrites, cinnabar. Bowles, 12. Berolding, 39.

In Rubbleſtone.

Copper pyrites, ſulphur pyrites, galena, biſmuthic and cobaltic ores.

In Clay.

In blue or grey clay pyrites moſt eaſily vitriolizable. Monnet Mineralog. 339.

In Gypſum.

Green calx of copper, copper pyrites, grey copper ore, galena, alſo ſulphur and bitumen.

In Strata of Coal. (Lehm. 219.)

Native ſilver, copper and ſulphur pyrites, galena, mangancſe.

Metallic Strata found in ſecondary Mountains. (Lehm. 220.)

Iron ſtone, ſulphur pyrites, calamine, cinnabar.

ESSAY X.

ON THE HUTTONIAN THEORY OF THE EARTH.

On this theory I have already made ſuch remarks as appeared to me ſufficient to ſhew its inconſiſtency with actual appearances, in a memoir inſerted among thoſe of the Royal Iriſh Academy, in the fifth volume; to theſe Dr. Hutton has ſince thought proper to reply with much acrimony, in an enlarged republication of his theory, forming two thick 8vo volumes; a detailed examination of all he has there advanced would neceſſarily be as voluminous; but, if I miſtake not, wholly ſuperfluous. It will be ſufficient to ſubvert the fundamental principles upon which his ſyſtem is conſtructed, and occaſionally to point out the abſurdities that flow from it: to fulfil this taſk we need only examine a few chapters of his work.

In his firſt chapter, he endeavours to eſtabliſh the fundamental points of his theory, which may be comprized in the following propoſitions:

1°. That the ſoil which ſerves as a baſis for vegetation is nothing but the materials collected from the deſtruction of the ſolid land, and that it, and the materials of which mountains are formed, are gradually detached, and unceaſingly carried forwards by the circulation and propulſion of water into the unfathomable regions of the ſea, and, conſequently, that in ſome indefinite time the land muſt be deſtroyed; but is afterwards to be renovated by ſubſequent operations; theſe operations, which lead partly to deſtruction and partly to renovation, he calls the *natural operations of the globe.*

2°. That the dry land which we now inhabit, was formed in the ocean, from materials conveyed into it, from an anterior habitable earth, and in particular that calcareous earth, or limeſtone, originates from marine animals.

3°. That the looſe materials thus collected, muſt have been conſolidated by heat and congelation from a previous fuſion, and

not

not by accretion, or cryſtallization from water.

Laſtly, that the ſubſtances thus conſolidated, have been elevated to their actual height over the ſeas and ocean, by extreme heat, and expanded by an amazing force, and form our preſent continents.

Examination of the firſt Propoſition.

This conſiſts of ſeveral parts; 1°. That all our actual ſoil, or the actual ſurface of the earth, conſiſts of particles detached from the more ſolid land, and that all mountains are in a ſtate of increaſing decompoſition. 2°. That this ſoil, and thoſe particles, detached by decompoſition, are neceſſarily carried into the abyſſes of the ocean. 3°. That in conſequence of this conſtant decay and deſtruction, all our mountains muſt in time be levelled, and our continents deſtroyed.

To juſtify this concluſion, the two firſt mentioned facts ſhould be eſtabliſhed in their univerſality; for if all *ſoil* does not ariſe from decompoſition, and if all *mountains* are not in a ſtate of decay, there is no

 room

room to ſuppoſe that the former conſtantly travels to the ſea, nor that all the latter will ever be levelled. Now all ſoil does not ariſe from decompoſition, for a calcareous ſoil often covers a clay that contains no limeſtone, and the ſolid rock under this, is often of the granitic kind; of this I have already quoted an inſtance obſerved at Dolau in Germany, where an effervеſcent calciſerous clay covers another which does not effervеſce, and conſequently of a totally different nature. Sauſſure alſo remarks that the ſand near Aliaſſon appears not to have been abraded from any rock, but to have cryſtallized on the ſpot whereon it is depoſited. 5 Sauſſ. § 1375. Therefore theſe ſoils did not ariſe from the decompoſition of the ſtones at leaſt, on which they reſt; nor are all mountains ſubject to decay, for inſtance, ſcarce any of thoſe that conſiſt of red granite. The ſtone of which the runic rocks are formed have withſtood decompoſition theſe 2000 years, as their characters evince, and of this ſort of granite whole mountains are formed in Sweden and Finland. 19 Schwed. Abhand. 221. 223, 224. Patrin alſo remarks the ſame indeſtructibility

lity of granites near Lyons. 4 Nev. Nord. Beytr. 170. This univerſal degradation of mountains is alſo denied by that great obſerver, Mr. Monnet; ſee his Mineralogy, p. 61, in the note. Baſaltic pillars in general bid defiance to decay, as is evident from their angles; even where this degradation takes place, it is in a degree conſtantly decreaſing, and, conſequently, muſt ceaſe long before the period at which the mountains could be levelled with the plains, either through the protection of vegetation, as Mr. de Luc has amply ſhewn in his 28th and 29th, &c. Letters to the Queen, or through other cauſes; for the rocks that ſtood at the entrance of the port of Alexandria, though continually beaten by the waves theſe 2000 years, ſtill ſubſiſt in their ancient integrity, as Dolomieu has well remarked. 42 Roz. 50. However I ſhall readily allow that in many places much of the ſoil has ariſen from the decompoſition of the ſtone whereon it reſts, or was conveyed by the ancient extenſion of rivers or floods from circumjacent mountains; but I muſt deny that it is in general true that it travels, ever ſo ſlowly, to the ſea, as where

 there

there is no declivity, there is no reaſon to ſuppoſe that either it or the water that falls upon it are carried forward; and thus, by far the greater part of the earth's ſurface is circumſtanced. Even in caſes where the progreſs of detached particles has been at firſt forwarded by declivities, it is afterwards frequently arreſted by the expanſion of the waters that convey them. For inſtance, in lakes where theſe particles are for the moſt part depoſited; hence Mr. Aikin could diſcover ſcarce any earthy particles in the waters of the lake Bala, in the midſt of Wales. See his Tour, p. 24. and as to ſtones of any bulk, Bowles has ſhewn, that none, not even of thoſe that have been rounded by rolling, arrive at the ſea. Bowles *Spain*. 506, &c.

But the moſt important miſtake in this part of the Doctor's Theory conſiſts in his ſuppoſition, that the ſand and ſoil conveyed to the ſea, by many and even moſt, great rivers, are depoſited in the unfathomable depths of the ocean; whereas nothing is more certain than that they are depoſited at the mouths of thoſe rivers, or at a ſmall diſtance from thoſe mouths, and there form prolongations of the land that incroach on the

the ſea; among innumerable inſtances, I ſhall quote only a few. Notwithſtanding the rapidity of the Rhone, it was from its depoſitions that the plains of Crau and Camarque, in lower Languedoc, were formed, and what is carried into the ſea is rejected on the ſhore by the currents of the Mediterranean, as Mr. Pouget has ſhewn. Mem. Par. 1775, 562. Dolomieu has proved that great part of lower Lombardy has been won from the ſea by the depoſitions of the Po; and inſtead of the ſea's retreating from Ravenna, as many have imagined, that the land has been extended by the ſame cauſe, much beyond that city. 42 Roz. 47, &c. All the ancients agree that the greater part of Egypt originated from the depoſitions of the Nile, and at this day, the ſludge depoſited at its mouth forms ſhoals which are elevated above the level of the river by the waves of the ſea which repel the ſand and mud, as Volney obſerved. Egypt, p. 31. and hence the old town of Damietta, anciently ſituated near the ſea, is now by the prolongation of the land, 2 leagues from it. 42 Roz. 196.— In the ſame manner Holland, and great part of Flanders, and the ſhoals on their coaſts, originated from the depoſitions of the Rhine, Meuſe, Schelde, &c. as is generally acknowledged, and is evident from the fluviatile ſhells diſcovered in ſundry excavations. The greater and more rapid rivers, as the Ganges, and Bourampooter, after depoſiting the greater part of the ſand and mud with which they are charged, at their mouths, and forming extenſive Deltas, convey ſome part into the ſea, but to no great diſtance, as Major Rennell has obſerved. Hence there is no reaſon whatſoever to ſuppoſe that inland waters convey any portion of earth into the unfathomable depths of the ſea, but there are many that may perſuade us, that not a particle is carried to any conſiderable diſtance from the coaſts. Mariners were accuſtomed for ſome centuries back to diſcover their ſituation by the kind of earth or ſand brought up by their ſounding plummets, a method which would prove fallacious if the ſurface of the bottom did not continue invariably the ſame. Fortis in his travels through Dalmatia, p. 282, relates that urns thrown into the Adriatic upwards of 1400 years, ſo

far

ſar from being covered with mud, were found in the ſame ſituation as they could be ſuppoſed to have been the firſt day of their fall; therefore notwithſtanding many particles of earth are by rivers conducted to the ſea, yet none are conveyed to any diſtance, but are either depoſited at their mouths, or rejected by currents or by tides*. And the reaſon is, becauſe the tide of flood is always more impetuous and forcible than the tide of ebb, the advancing waves being preſſed forward by the countleſs number behind them, whereas the retreating are preſſed backward by a far ſmaller number, as muſt be evident to an attentive ſpectator, and hence it is that all floating things caſt into the ſea, are at laſt thrown on ſhore, and not conveyed into the mid regions of the ſea, as they ſhould be if the reciprocal undulations of the tides were equally powerful. Friſi in ſome of his mathematical treatiſes remarks, that if any conſiderable maſs of matter were accumulated in the interior of the ocean, the diurnal motion of the globe would be diſturbed, and conſe-

* See 1 Varen. chap. 18, prep. 9.

quently it would be perceptible, a phenomenon, however, of which no hiſtory or tradition gives any account. Polybius, ſeduced by the ſame reaſons as Dr. Hutton, imagined that the Black Sea, which is ſo narrow, and into which ſo many great rivers diſcharge themſelves, would in ſome centuries be filled up, and by computation it may be ſhewn it ſhould. See 2 Buff. Hiſt. 70, &c. and 154. and yet after more than 20 centuries that have ſince elapſed, we have no reaſon to think that it is at preſent ſhallower than in his time, on the contrary, its ſurface has been narrowed by immenſe prolongations of the continent, and in this manner, if the rivers did not gradually decreaſe, it would probably at laſt be nearly filled, but not by any elevations in the middle (which would be the conſequence of the Doctor's theory), of which there is not the leaſt appearance. So the whole country about Peterſburg was formed by depoſitions from the Neva, and even ſtrata of ſtone were thus formed, for a boat, and human ribs, and reeds, were found under the ſtony ſtratum. 1 Nev. Nord. Betyr. 133. Here perhaps we might ſtop, as this principal

principal ſoundation of the Doctor's theory being ſubverted, it were needleſs to detect its remaining defects; but as he may reply, that if only one particle were detached from the moſt ſolid rocks in 4000 years, and only one particle carried into the mid regions of the deep in the ſame ſpace of time (a ſuppoſition it would be difficult to diſprove), in ſome immenſe future period the ſolid land of our continent would be deſtroyed, a cataſtrophe which he imagines has already ſeveral times happened, it will be neceſſary to ſhew from the inconſiſtency of his hypotheſis with actual appearances, that whatever may happen in future, nothing of this ſort has heretofore happened.

Examination of the ſecond Propoſition.

That the dry land which we now inhabit was formed in the boſom of the ocean, I have already proved in the former eſſays; though not from the materials of a former habitable globe, of which the Doctor brings no proof but what has been juſt refuted; but that all calcareous earth originates from ſea animals, I muſt deny; on the contrary, we have ſtrong reaſon to think that marine animals obtain calcareous earth from their food, for it is well known that ſea water contains ſelenite; and with reſpect to land animals, it is ſaid that hens produce eggs without ſhells when deprived of acceſs to mortar or other calcareous matter; whether this be true or not, it is certain that all animals and moſt vegetables, take in calcareous earth with their food, of which it appears to be an eſſential ingredient, therefore, an *habitable* earth ſimilar to ours, (and ſuch Dr. Hutton ſuppoſes) could not have exiſted at any period without that earth. But whether *habitable* or not *habitable*, unleſs the Doctor will alſo ſuppoſe the globe to have exiſted without any ſtone, he cannot conſiſtently with actual appearances ſuppoſe it to have exiſted without this earth, as there is no ſpecies of ſtone (except a few of the rareſt) into whoſe compoſition it does not enter, conſequently it muſt have exiſted in thoſe that formed the baſis of the ſea, and therefore have preceded the exiſtence of ſea animals. The Doctor indeed allows that vaſt maſſes of calcareous matter exiſt in which no animal remains can be diſcovered,

diſcovered, but theſe, he ſays, were cryſtallized by fuſion, which obliterated all trace of animal matter, p. 322, 325. the truth of this aſſertion we ſhall now examine.

Examination of the third Propoſition.

To decide the queſtion how ſuch continents as we now have, could have been elevated above the ſea, from materials collected at its bottom (as he thinks he has proved in his firſt propoſition) we muſt, he ſays, examine how they could have been conſolidated into maſſes of the greateſt ſolidity having neither water nor vacuity between their various conſtituent parts, nor in the pores of theſe conſtituent parts themſelves; but we have already ſhewn that there is no reaſon to think that our continents have ariſen from the materials collected at the bottom of the ſea, from the *detritus* of a former continent, nor is it true that minerals contain no water in their compoſition or between their pores but the very reverſe is the truth as we ſhall preſently ſee. There are but two ways, the Doctor

Doctor tells us, by which ſpungy bodies may be conſolidated and ſubſtances formed into maſſes of a regular ſtructure; the one of theſe is *congelation* from a fluid ſtate by cold; the other by *accretion* from a ſolution in water, p. 44. This laſt mode he rejects for reaſons which I ſhall ſoon examine, and adopts the firſt, principally from the alleged impoſſibility of applying the ſecond. The only direct reaſons he adduces are the following.

1° Flints are found perfectly inſulated both in chalk and ſand, and their form demonſtrates that they have been introduced by injection, and have congealed from a ſtate of fuſion; but to deſcribe theſe convincing appearances, would, he ſays, require a too prolix detail; inſpection alone he thinks ſufficient; yet to Mr. De Luc and other geologiſts, inſpection, it ſeems, has ſuggeſted a very different theory. This ſort of injection, of whoſe ingreſs not the leaſt trace is to be found, ſeems to me impoſſible; it is to be noted that flints are found in ſtrata, and at different heights, and placed with the utmoſt regularity, appearances utterly incompatible with this theory.

theory. Many other ſolid objections to it are urged by Mr. De Luc, in the Monthly Review for 1790, p. 209, to which I refer, but a demonſtration of its falſehood has lately been diſcovered, for 126 ſilver coins have been found incloſed in flints, at Grinoc, in Denmark. Schneider Topograph. Mineral. 114. and an iron nail at Potſdam, Ibid.

2° Pieces of foſſil wood have been found penetrated with flinty matters, in theſe the injected flint appears to him to to have been in a ſimple fluid ſtate, and not in a ſtate of ſolution in water, firſt, becauſe however little of the wood is left unpenetrated, the diviſion, he ſays, is always diſtinct between the injected part and that which is not penetrated by the fluid flint, there is no partial impregnation nor any gradation of the flintyfying operation, as muſt have been if the ſiliceous matter had been depoſited from a ſolution; and ſecondly becauſe the termination of the flinty impregnation has aſſumed ſuch a form, as would naturally happen from a fluid flint penetrating that body. Mr. De Luc, however, who has ſeen all theſe ſpecimens of foſſil wood, thinks their various appear-

ances

ances contradict the Doctor's hypothesis*, and indeed specimens occur whose appearances are absolutely incompatible with it; for instance, those in which the ligneous fibres are distinctly petrified, so that each of them may be separated from the other, and those which exactly exhibit the traces of organization. Could these arise from a mere fiery injection that should destroy all *fibres*, and form an uniform mass†? and in every case what should become of the ligneous part displaced by the injection? it could not evaporate or be burned, for the Doctor supposes a compression which should prevent all evaporation, and it was not driven out, for the place it passed through should be discerned, but this is one of the many inconsistencies of his system which the Doctor has entirely overlooked. In some cases, only the worms that corroded the wood are petrified. What confined his melted flint to these only? See Neret's paper in 17 Roz. Whereas that

* Monthly Review, 1790, p. 211.

† 17 Roz. 303, &c. and 6 Nev. Nord. Beytrage 118. and Fougeroux in Mem. Par. 1759, quoted in 10 Buff. Sup. 191.

water

water has frequently a petrifying power is notorious and evident, as the petrifaction of wood inserted in such water reaches as high as the water reaches, and no higher, and the water is found impregnated with the same stony matter as that of the petrifaction. See the instance recorded at the end of the first volume of Don Ulloa's Voyages, and the petrifaction of the wood of Trajan's Bridge, mentioned in the 4th Essay.

That petrifactions proceed from water impregnated with the petrifying matter appears also from this, that wood petrified by iron is more plentifully found in the vicinity of martial springs. 1 Nev. Nord. Betyr. 135. neither is the first fact on which the Doctor principally relies generally true. Collini in the 12th chapter of his Voyages, p. 166, French edition, tells us that in some shells converted into agates, the siliceous impregnation is *partial* and *gradual.* J'ai observé que dans un même noyau surtout de turbinite, les degrés de lapidification varioient de maniere, qu'il etoit totalement converti en agate transparente au sommet et qu' à mesure qu'on remontoit vers l'ouverture de la coquille la lapidification n'avoit produit qu'une pierre moins dure et opaque, jusqu' à la partie superieure du noyau qui n'avoit acquis qu'une tres legere cohesion.

3°. Fossil alkali called trona has been found in the southern parts of Africa, crystallized in a peculiar manner, which, contrary to the habit of common soda, does not effervesce nor liquify in a moderate heat, and as the Doctor thinks, contains no water of crystallization, and therefore must have been in a state of fusion immediately before its congelation and crystallization. If it has been fused, it must have been so in circumstances very different from those in which the Doctor supposes such extraordinary fusions to have taken place, for it is found not running in veins in beds of rock salt, as Doctor Monro supposed, but on the surface of the earth*, forming a crust at most an inch thick, consequently, if fused, most of its fixed air must have escaped, whereas it contains somewhat more of this than common soda does.

* 35 Schwed Abhandl. 131.

Through

Through the goodneſs of Dr. Black, I obtained nearly an ounce of this ſubſtance, and found its internal ſtructure nearly as mentioned; it did not effervefce in a ſlight heat as common ſoda does, but it did effervefce in a heat of 300°, and therefore loſt water; but to determine the quantity of this water, I ſaturated the trona with vitriolic acid, and found it to contain 34 per cent nearly of fixed air, 41,9 of mere alkali, and 1,8 of reddiſh ſand, the remaining 22,3 grains were therefore water. This proportion of water is indeed much ſmaller than that contained in 100 parts of common ſoda, and the proportion of fixed air is greater; but this ariſes from the different circumſtances of its cryſtallization. Common ſoda is cryſtallized in mere water, this on the contrary germinated from the earth, and was expoſed to a burning ſun; and it is well known that ſalts cryſtallized at a high temperature contain leſs water than thoſe in a low. It is evident however, it was not fuſed, elſe the few grains of ſand found in it would have been vitrified. His remaining proofs being indirect, conſiſting in objections to the theory of

 the

the aqueous ſolution of minerals, I ſhall conſider under that head: as for his arguments from the ſtructure of *ſeptaria*, I do not perceive their force, and therefore paſſing them by, I ſhall now ſtate my reaſons for rejecting the opinion that minerals owe their ſolidity to congelation from fuſion.

'My firſt objection to this theory is, that it is grounded on a ſuppoſition (or at leaſt we may infer from it) that at ſome paſt period a degree of heat prevailed under the ocean ſuperior to any that has ever been known to exiſt, and which muſt have taken effect in circumſtances the leaſt favourable to its production and exertion; thus in the Doctor's hypotheſis, the enormous mountains of calcareous matter in which no animal remains have been detected, ſuch as are found in Swiſſerland, Tyrole, Siberia, &c. were fuſed, and not only fuſed but cryſtallized by this heat: to form ſome idea of the heat requiſite to effect this fuſion, we muſt recollect that the heat produced in the focus of a large burning glaſs, is much ſuperior to any that can be produced in any furnace, but much inferior to that produced by pure air acting on burning charcoal, and yet neither Lavoiſier nor any other experimenter has been able to fuſe the ſmalleſt viſible particle of pure limeſtone in this prodigious heat. Mr. Sauſſure indeed, has of late ſucceeded in melting a particle of it, but ſo ſmall that it could be diſcerned only by a microſcope. What then muſt have been the heat neceſſary to melt whole mountains of this matter? Judging then from all we at preſent know of heat, ſuch a high degree could only be produced by the pureſt air acting on an enormous quantity of combuſtible matter: now Ehrman obſerved that the combuſtion of 280 cubic inches of air acting on charcoal, was not able to effect the fuſion of one grain of Carrara marble, from whence it is apparent that all the air in the atmoſphere, nor in ten atmoſpheres, would not melt a ſingle mountain of this ſubſtance of any extent, even if there were a ſufficient quantity of inflammable matter for it to act upon. Judging alſo of ſubterraneous heat by what we know of that of volcanoes, no ſuch heat exiſts; the higheſt they in general produce is that requiſite for the fuſion of the volcanic

 glaſs,

glaſs, called obſidian, which Sauſſure found not to exceed 115° of Wedgewood, but baſaltine, which requires 140° of Wedgewood, is never melted in the lavas of Etna; how little capable then would volcanic heat be to effect the fuſion of Carrara marble, which, according to the ſame excellent author, would require a heat of upwards of 6300° of Wedgewood, if this pyrometer could extend ſo far? And in what circumſtances does Doctor Hutton ſuppoſe this aſtoniſhing heat to have exiſted, and even ſtill to exiſt? Under the ocean in the bowels of the earth, where neither a ſufficient quantity of pure air nor of combuſtible matter capable of ſuch mighty effects can with any appearance of probability be ſuppoſed to exiſt; and without theſe, ſuch degrees of heat cannot even be imagined without flying into the region of chimeras. Volcanoes give no countenance to ſuch a ſuppoſition, their heat is not commonly greater than that which may be effected by the abſorption of ſulphur by various metallic ſubſtances, as the Dutch philoſophers have lately ſhewn. To this train of reaſoning Doctor Hutton anſwers, p. 253, "That it

"was

" was improper to try his theory by fire." And p. 35, " There may be ſome difficulty " in conceiving the modifications of ſubter- " raneous heat; but as on the one hand we " are not arbitrarily to aſſume an agent for " the purpoſe of explaining events, or cer- " tain appearances, which are not under- " ſtood, ſo on the other we muſt not re- " fuſe to admit the action of a *known* pow- " er, *when this is properly ſuggeſted by ap-* " *pearances*; and though we may not un- " derſtand the modifications, capacity, and " regulation of this power, we are not to " neglect the appropriating to it as a cauſe, " *thoſe effects that are natural to it*, and " which, ſo far as we know, cannot belong " to any other." Much is here untruly ſuppoſed with reſpect to the *known* power of heat, which I ſhall not now notice, but ſingly examine how far its operation is *ſug-geſted* by the actual ſtate and *appearance* of mineral ſubſtances.

My next objection then to the Doctor's theory is, that the actual ſtate of all minerals preſents appearances incompatible with the ſuppoſition that they originated from heat or fuſion. 1°. Almoſt all ſtones loſe ſome

 part

part of their weight *, and are altered either in their hardneſs, luſtre, colour, or permeability, to light, when heated; could this happen if to heat they owed their origin? the aſſiſtance of compreſſion in this and the following inſtances is in vain invoked.

The ingredients of granite are found diſtinct and ſeparate from each other, and the felſpar opake and without blebs; now if theſe ingredients were fuſed they would run into each other, and the felſpar would either have blebs, or if melted in a high heat would form a tranſparent glaſs; the quartz would alſo become opake, as it always does on being heated, the mica alſo would have been vitrified, therefore the *appearance* of this aggregate is incompatible with the ſuppoſition of its having been fuſed. Sir James Hall attempts an anſwer to this objection, which I ſhall examine hereafter.

2°. Granularly foliated limeſtone, in which no animal remains exiſt, and which the Doctor would perſuade us was cryſtallized by fuſion, is often found ſeated on

* See Eſſay IV.

argillite,

argillite, and even intercepted between the laminæ of argillite; yet the argillite, (one of the moſt fuſible of all ſtones,) not only diſcovers no marks of fuſion, nor even of having been heated; Is not this impoſſible on the ſuppoſition that the limeſtone was melted? The Doctor tells us, that ſhells originally exiſted in limeſtone, but were deſtroyed by heat; but here he contradicts his own hypotheſis, namely, that the fuſion took effect under a compreſſion that prevented the volatilization of all aerial fluids; if he ſays they remain, but unaltered in their form, I aſk, how they came to remain unaltered in their form in a heat ſufficient to melt them, as they do in many ſiliceous ſtones? for in company with theſe, which according to him were in fuſion, *they* alſo ſhould have been fuſed. Again, ſtrata of granular limeſtone are often intercepted between ſtrata of gneiſs, and even alternate with them, and metallic veins paſs without interruption through each, a proof of their coeval formation; but where the vein paſſes through the limeſtone it is deſtitute of metal; and even the metallic ſubſtances found in each differ both in nature and poſition, as Charpentier

has

has obſerved. *Lettere Oritologice al Sr Arduino.* Now it is evident that the heat which could melt the limeſtone ſhould melt alſo the ſtrata of gneiſs, which is much more fuſible, and confound both. This obſervation alſo proves that metals were not thrown up in a ſtate of fuſion into the veins that contain them (another opinion of Dr. Hutton's), elſe ſome would be found in that part of the vein that traverſes the limeſtone.

To theſe conſiderations I ſhall add a demonſtration that pure limeſtone, ſuch as Carrara marble, did not originate from ſhells; it is this, phoſphoric acid is found in moſt ſhells *, but none is found in pure limeſtone, Mem. Turin. 1789, 63. and its abſence cannot be attributed to fuſion, for phoſphoric acid is indeſtructible by heat.

3°. Stony ſubſtances are often found ſuperimpoſed on clay, which has not the appearance even of having been baked; how could that happen if both were heated from beneath?

4°. Various ſiliceous cryſtals, and ſome gems are found terminated by perfect py-

* See Mr. Hatchett's curious paper in Phil. Tranſ. 1799, Part II.

ramids

ramids at both ends, without any mark of adherence; if these were formed in water such formation may be conceived, as the primordial particles might be supported by that fluid, but they cannot be supposed to have been formed by fusion, as during that operation they must have been supported by some solid to which they would afterwards adhere, and thus their shape on separation would be destroyed.

5°. The impressions of shorls and other stones more fusible than quartz, are often found deeply engraved in it; they must therefore, if in fusion, have congealed, while the quartz was still in fusion which is a contradiction in terms.

6°. Pure minute siliceous crystals are found dispersed even in Carrara marble; the granular limestone of Dauria and Swisserland abound in siliceous and argillaceous particles; now if these calcareous masses had been fused, this could not happen, as these substances are fluxes to each other, and run into a common mass. Fluors and gypsum have also been found crystallized in the midst of substances into which, if fused, they should flow, and with them, by the

the laws of affinity, form one common maſs, as is daily experienced.

7°. Shells, in their natural ſhape and with their natural pearly luſtre, have been found imbedded in flints and hornſtones; Could this happen if the flints, &c. had been melted? if it could, why are they never found in granite or gneiſs, &c? Again, as to metallic ſubſtances and ores, they equally diſclaim an igneous origin.

Thus, 1°. Gold is found native in large maſſes in the county of Wicklow, many ſpecimens, to all external appearance even when examined with a microſcope, perfectly free from ſtony ſubſtances. I have examined one of theſe ſpecimens, and found its ſpecific gravity only 13. but after it had been melted its ſpecific gravity amounted to 18. and many ſandy particles appeared on its ſurface; theſe, therefore (and perhaps ſome vacuities), were originally diſperſed through its maſs, which could not happen if it had originated from fuſion, ſince by the laws of gravity the grains of ſand ſhould float on its ſurface, as they afterwards did when it was really fuſed. Mr. Alcohorn, I am informed, ſawed a piece of this found

in

in Wicklow, and found grains of quartz and ironſtone in the middle of it. In the ſame manner native ſilver is often found internally mixed with quartz, 2 Roz. 197, which could not happen if the ſilver had been fuſed.

2°. The ſulphurated ſilver ore is frequently found incorporated with calx of iron; now if it had been formed by fuſion, this could not happen, as the ſulphur would unite preferably to the iron.

3°. Cryſtallized amalgamas of ſilver and mercury are found in the Dutchy of Deux Ponts. Could theſe have been formed by heat? Magnetiſm is deſtroyed by heat, and magnets cannot be formed by fuſion, but they may by ruſt and water, therefore that iron ore was never fuſed.

3°. Specular iron is found lining the inſide of gun barrels in which water had been decompoſed, by fuſion this compound would be deſtroyed; the natural was not therefore formed by fuſion.

5°. Galena may be imitated in the dry way, but imperfectly, for that ſo formed does not decrepitate when heated, whereas the natural does, therefore this was not formed by fuſion.

6°. Pyrites

6°. Pyrites may be formed by fusion, but such pyrites differs much from the natural, the natural is yellow, the artificial never; the natural is scarcely fusible, the artificial easily; the natural contains abundance of sulphur and yields it by distillation; artificial cannot be made to yield any.

7°. Cinnabar may also be formed by heat, but the artificial differs from the native; the native is almost always of a compact or foliated texture, whereas that formed by heat is fibrous or striated.

8°. So also regulus of antimony sulphurated by nature takes up upwards of 35 per cent. of sulphur, but by fusion the same quantity of regulus cannot be made to take up 30 per cent. of sulphur.

9°. Blendes, or sulphurated zinc ores, contain a large proportion of water, which could not happen if they were produced by fusion, but is not surprising if they were formed in water; moreover sulphur and zinc can scarcely be brought to unite in the dry way.

10°. Native precipitate, *per se*, or oxygenated mercury, is frequently diffused through bituminous clay or slates; heat would immediately ſeparate the oxygen, this ore therefore was formed in the moiſt way.

11°. Cobalt and biſmuth are found united in cobaltic ores, but theſe metals do not mix with each other in fuſion, therefore they were not formed by it.

12°. In the cobalt ore, called *reticular cobalt*, we ſee the cobalt running in veins through quartz, and yet the quartz on each ſide remains untinged, which could not happen if the cobalt had been in fuſion. In many caſes we ſee the quartz tinged *red* by cobalt, whereas if fuſion had taken place, the tinge ſhould have been *blue*.

13°. White calx of mangaueſe blackens when heated, now this calx is found native, therefore this was never heated.

14°. Sulphur if heated in contact with oxygen, muſt have been converted into vitriolic acid; and for the ſame reaſon, no native regulus, but rather calx of arſenic, ſhould exiſt, therefore the *abſence* of oxygen in a heat that could be produced only *by oxygen*, is a new ſuppoſition to be added to the Huttonian theory.

Theſe inſtances I think ſufficient to

prove,

prove, that the actual *state and appearance* of minerals are not due to previous fusion.

The Doctor indeed, seems to rest the necessity of admitting his theory chiefly on the impossibility of ascribing the acknowledged original fluidity or softness of minerals to any other cause but igneous fusion. There are just two ways, he says, by which porous bodies may be consolidated, and formed into masses of a regular structure, namely, congelation from a fluid state by cold, and accretion by crystallization from water. His arguments against this last widely dispersed through his whole work, or at least, the principal among them, I shall now consider, but expressed in my own words, as the author's statements are intolerably perplexed and diffuse, and moreover frequently refer to that part of his theory which I have already refuted, namely, that all minerals were originally collected at the bottom of the sea from the materials of an anterior continent, and consequently in a *spungy porous* state, as he calls it.

1°. If it was from a solution in water that minerals were formed, it is impossible

to

to conceive how theſe maſſes ſhould have been conſolidated without any viſible water in their compoſition; the anſwer to this is obvious; ſalts daily cryſtallize in water, without retaining any *viſible* in their compoſition. Barytic lime acquires a ſtony hardneſs on being barely ſlacked *. The depoſitions of the ſprings of Cartſbad, in Bohemia, the cruſt formed in common tea kettles are other inſtances of ſtony maſſes formed in water without retaining any viſible in their compoſition. Doctor Black can tell him, " that when concretions of ſiliceous earth are once formed, and afterwards receive frequent additions of the ſame matter, which inſinuating itſelf into the pores of the concretion is fixed there, and increaſes their denſity and ſolidity, the maſs may in time acquire a ſurpriſing degree of hardneſs, the petrifactions of Geyſer are undoubtedly formed in this manner, and ſome of them are ſo denſe and hard, that they can hardly be diſtinguiſhed from agate or calcedony. 3 Edin. Tranſ. p. 114. .The obſervations of

* Annales de Chym. p. 278, 283.

 Rinman,

Rinman, and Mr. Edward King, quoted at large in my 4th Eſſay, are alſo full to the purpoſe.

2°. The Doctor inſiſts, that if water had been the menſtruum by which conſolidating matter had been introduced into the interſtices of ſtrata, then water ſhould be conſidered as an univerſal menſtruum, in contradiction to chemical principles, for there are ſtrata conſolidated by calcareous ſpar, by fluors, by ſulphureous or bituminous ſubſtances, by ſiliceous matter; all which ſubſtances are inſoluble in water. P. 53. he aſſerts that no ſiliceous body having the hardneſs of flint, nor any cryſtallization of that ſubſtance, have ever been formed except by fuſion. From this paragraph one would be apt to infer that flints and ſiliceous cryſtals have in ſome experiments been formed by fuſion, whereas the truth is, that no flint or pure ſiliceous cryſtal have ever been known to have been fuſed, except the microſcopic particle fuſed by Sauſſure, as already mentioned, and that the fuſibility of ſiliceous matter in any known heat, without the aſſiſtance of fluxes, which enter into the compoſition of the reſulting compound,

pound, is as much or more oppoſite to chemical principles than the ſolubility of ſiliceous matter in water, I ſay more oppoſite, becauſe ſiliceous particles have frequently been found diſſolved in water, as I have ſhewn in many inſtances, both in the 4th and 7th Eſſays, and therefore need not here repeat. The inſolubility of ſiliceous matter proceeds from its integrant affinity: thus Mr. Macie found powdered flints inſoluble. Phil. Tranſ. 1791. 385. hence when it is ſufficiently comminuted, as I have ſhewn it originally to have been, there is no obſtacle to its ſolution. Water cannot, however, be called an univerſal menſtruum, as a menſtruum is a fluid whoſe ſpecific affinity to the particles of a ſolvend is greater than the integrant affinity of the ultimate particles of the ſolvend to each other. His aſſertion that ſtrata are ſolidified by ſpars, ſiliceous matter, fluors, &c. is perfectly chimerical, and reſts ſolely on his own peculiar opinion that they were originally depoſited in a ſoft ſtate, but hardened by the congelation of thoſe fuſed ſubſtances; whereas their hardneſs proceeds generally, merely from their own integrant attraction, as ſhewn in the 4th Eſſay. From what elſe can the hardneſs even of a ſubſtance congealed after fuſion proceed ?

3°. Sulphur is found combined with almoſt all metals; a combination, he ſays, that could be formed only by fuſion, as ſulphur is inſoluble in water. This aſſertion alſo is contradicted by facts: ſulphur has been repeatedly found diſſolved in water in the form of hepatic air; I have ſhewn that a cubic foot of water may contain 374 grains of ſulphur, and according to Bergman, it may contain much more. See 2 Weſtrumb. 117, 130. Copper pyrites have been found newly produced, as ſhewn in my 7th Eſſay, and alſo the red ſilver ore.

Monnet aſſerts martial pyrites to be daily formed, Vitriolization, p. 11. he alſo found that calces of iron and ſulphur unite in the moiſt way. Diſſol. p. 57. That copper and ſulphur unite in the moiſt way, See 1 Chy. Ann. 1794, 296. Blende is unluckily quoted by the Doctor, as from the difficulty of uniting ſulphur and zink in the dry way, and the quantity of water it contains it *prima facie* beſpeaks an aqueous origin. Cinnabar and antimony are alſo

This account of the origin of mineral coal appears to me as improbable as that of the origin of ſtony ſubſtances, and in ſome reſpects, more extraordinary, as will appear by examining it in detail.

1°. According to this hypotheſis, all fuliginoſities ariſing from combuſtion on the ſurface of the earth are finally carried into the ſea; this reſts on the ſuppoſition, that all ſoil is gradually carried into the ſea, a notion which has been already refuted.

2°. It is ſuppoſed that all rivers carry vegetable carbonaceous matter into the ſea, and that it is there depoſited, and yet no proof is given that any river depoſits the vegetable matter that tinges them, and on the contrary it appears from the Doctor's own words, that this vegetable carbonaceous matter is depoſited, only in the caſe where the water is ſeparated from it by evaporation. If the action of the ſun and atmoſphere has the power of producing ſuch a ſeparation without evaporation, the Doctor ſhould adduce ſome experiment to that purpoſe, and not content himſelf with mere aſſertion; but the fact is, depoſitions of even ſlimy inflammable matter are exceeding rare, and muſt, like many others, be made in the rivers themſelves, as in the canals in Holland, or at their mouths, and there is no reaſon to think that it is the vegetable matter that diſcolours ſome rivers that is ever depoſited, but the more ſolid turfy particles that are carried off by the rivers that flow through bogs or the particles that accidently fall into them; the vegetable matter found in the canals of Holland does not form a particular ſtratum, but is mixed with mud, no ſuch ſtratum has been found in the excavation made in Holland to the depth of 240 feet, as may be ſeen in Varenius and Muſchenbroeck, nor in that made in the ſtrata formed by the depoſitions of the Seïne examined by Guettard, Mem. Par. 1753. nor in that made in Egypt, though the ſoil was intirely formed of the depoſitions from the Nile, nor has any ſuch coaly matter been ever brought up in any ſoundings; hence I conclude that the exiſtence of ſuch ſtrata at the bottom of the ſea is purely imaginary.

3. If our beds of mineral coal had been formed in the ſea, we ſhould find ſea ſhells among

among them, whereas fluviatile ſhells are by far the moſt common, when any are found. Sea ſhells occur very rarely, and theſe are merely adventitious, as I have ſhewn in the preceding Eſſays.

4°. Suppoſing even that vegetable matter had been conveyed into the ſea, and there had formed particular ſtrata, the diſtillation of theſe ſtrata is equally inconceivable; to diſtil coal a great heat is neceſſary; to diſtil it ſo as to expel the oily matter, an incandeſcent heat would be required, and even by this the laſt portions of bitumen cannot be expelled, moreover an immenſe quantity of inflammable air is produced: if our beds of coal had undergone that operation, would not various ſigns of it appear? Would not bitumen be found in the neighbourhood of thoſe beds of coal from which it had been expelled? Would not the ſulphur alſo be diſtilled from the pyrites found in the coal? Yet neither in the coal mines of Kilkenny, the coal of which is of all others moſt completely deſtitute of bituminous matter, nor any where near them, is the leaſt trace of bitumen

bitumen to be found, and the pyrites remain in their uſual integrity. Would not ſome of the neighbouring ſtones appear fuſed by this heat? Yet no trace of fuſion is ſeen in any coal mine. Would not the ſtrata of coal itſelf appear bloated and puffed, inſtead of aſſuming a regular foliated texture? Would not the ſuperior earthy or ſtony ſtrata be diſordered by the vehement expulſion of air? Yet no ſuch diſorder, but in general a great regularity, is obſerved. How comes it to paſs, that the few ſhells, and the by far more numerous leaves that are found in coal mines, diſcover not the leaſt mark of having endured any heat? And yet the interſtices of the coal are frequently lined with lamellar quartz, particularly in the coal mines of Kilkenny; a new proof that the quartz was never fuſed.

After theſe remarks, I think it needleſs to enlarge farther on this ſubject, or inquire why coal mines are confined to ſecondary hills, or inquire how fixed air had acquired its carbonic matter, and ſhall only add that the Doctor's aſſertion, that ſand is never found

found mixed in the ſtrata of coal, is not true. See Charp. p. 7. and 2 Buff. Miner. p. 189, in 8vo.

Even ſtrata of ſalt appear to the Doctor to have been formed by fuſion; but the only proof he adduces is, that the ſalt rock of Cheſhire lies in ſtrata of red marl, and that the regular ſtructure of the floating marly ſubſtance in the body of the ſalt is inexplicable on any other ſuppoſition but the fuſion of the ſalt. The intermixture of marl and clay in beds of ſalt, I have already explained, and its exiſtence in them is a certain proof that the ſalt was not fuſed, for if it were, it would by the mixture of theſe ſubſtances be decompoſed, at leaſt in part, whereas no trace of uncombined alkali is found in common ſalt mines; the ſea ſhells, and alſo wood, often found between the ſtrata diſcover no ſign of the application of heat; beſides ſalt was never in any experiment found cryſtallized after fuſion. Of metallic ores enough has been already ſaid; I ſhould not, however, forget in digging the ruins of the ancient town of Chatelet ſeveral iron tools were found, the wooden handles of which were converted

into

into hematites, the organization of the wood ſtill remaining; they lay buried during 1600 years. 10 Buffon Suppl. 197. This ore then muſt have originated in the moiſt way.

Having thus ſhewn that the origin and preſent ſtate of no ſpecies of mineral can, with any appearance of probability, be aſcribed to fuſion, I ſhould here conclude this Eſſay, as it were idle to inquire how our continents could be raiſed, by a cauſe of whoſe exiſtence we have no proof, did I not think it neceſſary to expoſe the fallacy of ſome, and the falſehood of others of Dr. Hutton's replies to the objections I made to his ſyſtem on a former occaſion. I ſhall not indeed always confine myſelf to his own words, as this would be often intolerably tedious, but I ſhall ſtrictly ſtate the ſubſtance of each.

P. 206. The Doctor denies that he had ſaid, " that *all* ſoil is made from the de-" compoſition or *detritus* of ſtony ſubſtan-" ces," yet page 13, he ſays, " a ſoil is *no-*" *thing* but the materials collected from " the deſtruction of the ſolid land;" this requires no comment.

Ibid.

Ibid. He denies that he had ſaid, " that " ſoil is *conſtantly* waſhed away, but only " that it is *neceſſarily* waſhed away, that is, " *occaſionally*." This is a mere play upon words, *conſtantly* is as often taken for *certainly* as *perpetually*, ſee Johnſon's Dictionary; but I am ſure *neceſſarily* was never uſed as ſynonymous to *occaſionally*.

Ibid. He apprehends " I have miſapprehended Mr. De Luc, when I aſſerted on his " authority, that ſoil is not always carried " away by water, even from mountains." The paſſages of Mr. De Luc are too long and too numerous to be here inſerted; they may be found in the 27th and 28th, and following letters to the Queen; on peruſing them it will eaſily be ſeen which of us has miſapprehended the meaning of that celebrated Geologiſt.

p. 208. He cenſures me for advancing, " that his concluſion relative to the imper- " fect conſtitution of the globe falls to the " ground, and the pains he takes to learn " by what means a decayed world may be " renovated are ſuperfluous."

In reply to which he aſſerts, " that the " object of his theory is to ſhew that this

" *decaying*

" *decaying nature* of the folid earth is the " very *perfection* of its conftitution as a *living* " world, and therefore it was proper he " fhould fhew how the decayed parts fhould " be renovated."

P. 210. He puts me a ftring of queftions. " Does he mean to fay, that it is " not the purpofe of this world to provide " foil for plants to grow in?" I anfwer, provifion for this purpofe has long ago been made and does not require daily renewal. " Does he fuppofe that foil is not remove- " able with the running water off the fur- " face?" I anfwer, that all that is move- able is not moved; all the water that falls on the furface does not run to the fea, but is either foaked, evaporated, or fucked in by vegetables; that which runs is often ftopped in its courfe, and the molecules of foil abraded and carried from fome fpots are often annually recruited by vegetation, except in a few particular fituations, as on the fides of hills much expofed to winds, and even there the abrafion is fcarce ever total. " Does he think it is not neceffary to re- " place the foil which is removed?" I anfwer, it is not, neither always nor every

 where.

where. The Doctor ſays, “ He required no “ more than this gradual removal of ſoil ;” but in the firſt place, he required this removal to be univerſal and complete, and in the next place, that this ſoil ſhould at laſt be ſeated in the unfathomable abyſſes of the ocean, notions which I have ſhewn to be unfounded, and he himſelf partly owns to be ſo in the note to page 14 of his new publication.

P. 211. The Doctor charges me with denying what he aſſerts in his theory, namely, “ That the ſolid parts of the globe are “ in general compoſed of ſand, gravel, ar- “ gillaceous and calcareous ſtrata,” and with adding, “ that this cannot be aſſumed “ as a fact, but rather the contrary; that it “ holds true only of the ſurface, and that “ the baſis of the greater part of Scotland “ is evidently a granitic rock, to ſay nothing “ of the continents both of the old and new “ world, according to the teſtimony of all “ mineralogiſts,” in anſwer to which, he tells us, that this general propoſition he ſtill maintains as a fact, and that after viſiting moſt parts of Scotland and obtaining good information with regard to thoſe

parts

parts which he had not ſeen, " that he can " with ſome confidence affirm, that (ex- " cept the north-weſt corner) inſtead of the " baſis of the greater part of Scotland being " a granitic rock, very little of it is ſo, not " perhaps $\frac{1}{300}$ part." His countryman, Mr. Williams, by profeſſion a miner, and who certainly has viſited all parts of Scotland, and particularly noted the ſubterraneous, gives us a very different account. Vol. II. p. 13. he tells us, " that the moun- " tains of Ben-Nevis in the Highlands are " chiefly compoſed of red granite, and that " it is found in *great abundance in many* " *other parts of Scotland* . . . . . but without " the leaſt appearance of ſtratification;" and p. 19, " That grey granite is very " common in many parts of this iſland both " in high and in low lands, north and " ſouth;" and p. 33, " that Scotland is re- " markable for a great number and variety " of granites."

Mr. Everſman, a German mineralogiſt, who reſided ſome years in Scotland, and viſited moſt parts of it, is ſtill more expreſs; he tells that the fundamental rock (Grundgebirge) conſiſts of a granitic aggregate

(grani-

nitic aggregate (granitartigen maſſe), 1 Berg. Jour. 1789, p. 495. The Doctor indeed tells us, that along the coaſt of Galloway to Inverary, he examined every ſpot between the Grampians and Tweedale mountains, yet could ſee no granite in its place; but Doctor Aſh in a letter to Mr. Crell, 1 Chy. Ann. 1792, p. 115, informs him, that from Galloway, Dumfries, and Berwick, there is a chain of mountains, commonly ſhiſtoſe, but often alſo granitic: and Mr. Grotſche, another German mineralogiſt, who had viſited Scotland, affirms, that the Grampian mountains conſiſt of micaceous limeſtone, gneiſs, porphyry, argillite, and granite, alternating with each other. 1 Bergb. 399. Moſt probably the Doctor examined only the ſurface, where he ſhould not always expect to meet that which I called the baſis the ſuperior ſtrata; the glaring inconſiſtencies that occur in page 215 I need not mention, as they cannot eſcape notice.

P. 216. He upbraids me with "forming "my notions of geology from the vague "opinion of others and not from what I "had ſeen." Muſt not many of thoſe who he expects will embrace his opinions, do the ſame? Does he think that from a view of Britain ſingly, a geology can be formed? Yet this is all he boaſts to have ſeen; though I have not travelled with my eyes ſhut, yet I felicitate myſelf with being acquainted, not merely with the opinions, but with the facts related by a Ferber, who has travelled through Germany, Italy, and England; by a Pallas, and Patrin, and Herman, who traverſed Ruſſia, and Siberia; by a Born, who viſited Hungary; by a Sauſſure, who made us ſo well acquainted with the Alps; by Carbonieres and La Perouſe, who viewed and examined the Pyrenees; by Charpentier, who inveſtigated ſo ably the internal ſtructure of Saxony; by Laſius, who diſplayed that of the Hartz; and by many more whom I have occaſionally quoted in the preceding Eſſays. I ſuppoſe it will be thought reaſonable to pay more attention to theſe, than (to expreſs myſelf in the mildeſt terms) to the *a priori* concluſions of any man, unſupported by facts, and contradicted by all natural appearances. As to the tendency of his ſyſtem to prove that this globe had properly ſpeaking no beginning, I ſhall take no far-

 ther

ther notice of it; all he ſays he means is, "that in tracing back the natural opera-"tions which have ſucceeded each other "... we come to a period in which we can-"not ſee any farther. This, however, is *not* "*the beginning* of thoſe operations which pro-"ceed in time .... nor is it the eſtabliſh-"ing of that which in the courſe of time "had no beginning, it is only the limit of "our retroſpective view of thoſe operations "which have come to paſs in time, and "have been conducted by ſupreme intelli-"gence." p. 223. Let the reader underſtand this as he can: but whether the Doctor can ſee ſo far or not, there muſt have been a primitive globe, or this globe was eternal; if there was a primitive globe, it muſt have had calcareous earth underived from ſhell fiſh, or not reſemble ours.

P. 226. As I had advanced in my former paper, that the interior parts of the earth, *at the depth of a few miles* might have been originally, as at preſent, a ſolid maſs, the Doctor aſks, "how a naturaliſt who "had ſeen a piece of Derbyſhire marble, or "any other limeſtone, could make that "ſuppoſition?" As if any marble or limeſtone had ever been taken from the depth of a few miles! and as if the ſea itſelf, from whence he derives all, did not require a ſolid body to ſupport it.

P. 227. He enters on a refutation of my notions of the conſolidation of ſtrata, which he contrives ſo to diſtort, miſtate, and perplex, that to diſentangle them from his diſingenuous miſrepreſentations would require too tedious a diſcuſſion; my own ſtatement may be ſeen in the preceding Eſſays, and differs but little from the ſentiments of ſome or other of the moſt enlightened naturaliſts of this age, Sauſſure, Werner, De Luc, La Metherie, &c. Yet as I had quoted the induration of *Pouzzolana mortar* (not common mortar as he miſtates) under water, he ſays, "One would imagine I was "writing to people of the laſt age," and takes no notice of the ſtalactite formed under water, which I had quoted from Mr. Smeaton's obſervations, though both are full proofs of the general fact, that ſtony concretions may be formed in water, though their interſtices were originally filled with water. What too will he ſay to Dolomieu's obſervation, that many ſtones harden by by ſprinkling them with water? Ponces, 417. and to many other inſtances which I have quoted in Eſſay IV? He ſurely cannot think that more regard is to be had to his *a priori* reaſoning about the difficulty of expelling water from the interſtices of concreting maſſes, than to known facts.

P. 236. After obſerving that I had taken great pains to refute the notion of a ſubterraneous heat ſufficient to melt all mineral ſubſtances, he tells us, " that he gives " himſelf very little trouble about that fire, " and takes no charge with regard to the " procuring of that power, as he had not " founded his theory on the *ſuppoſition* of " ſubterraneous fire; however that fire pro- " perly follows as a concluſion from thoſe " appearances on which the theory is found- " ed. . . . . he does not pretend to prove " demonſtrably that the fuſed minerals had " been even hot; however that concluſion " alſo naturally follows from their having " been in fuſion, it is ſufficient for him to " demonſtrate that theſe bodies muſt have " been more or leſs in a ſtate of ſoftneſs " and fluidity, without any ſpecies of ſolu- " tion; he does not ſay that this fuſion was

" without heat, but if it had, it would answer equally well the purpose of his " theory." And p. 237 he owns, " I have " justly remarked the difficulty of fire burning below the earth and sea, but says it " is not his purpose to endeavour to remove " those difficulties which perhaps only exist " in the suppositions made on that occasion. . . . . It is surely one thing to employ " fire and heat to melt mineral bodies, in " supposing this to be the cause of their " consolidation, and another thing to acknowledge fire or heat as having been " exerted on mineral bodies, when it is " proved by actual appearance that these " bodies had been in a melted state; here " are distinctions which would be thrown " away upon the vulgar," &c. I confess I am one of the vulgar, and can understand nothing in this paradoxical paragraph; I cannot conceive how minerals could be melted without heat, unless miraculously, nor the difference between the employment of heat, and the exertion of heat, and plainly see, that it would be idle in me to argue with a person who thinks he can.

P. 238. He tells us, " He does not avoid " meeting

" meeting the queſtion of providing the " materials for ſuch a mineral fire as may " be required; but it muſt not be put in " the manner I have put it, that is, as if he " had made that fire a neceſſary condition, " or principle of conſolidation, whereas he " had inferred the exiſtence of an internal " heat from the proofs he had given that " ſtony ſubſtances had been in a fluid ſtate " of fuſion, and if theſe be juſt then my " arguments are uſeleſs." And in p. 243, he tells us, " that according to his theory " the ſtrata of this earth are compoſed of " materials which came from a former " earth, particularly the combuſtible ſtrata " that contain plants. Let us then ſuppoſe " the ſubterraneous fire ſupplied with its " combuſtible materials from this ſource, " the vegetable bodies growing on the ſur-" face of the land. Here is a ſource pro-" vided for mineral fire which is inexhauſt-" ible, or unlimited, unleſs we circumſcribe " it with regard to time and the neceſſary " ingredients." As to the firſt part of this paragraph, I have ſufficiently refuted it by ſhewing that there is no general appearance, which neceſſarily ſuggeſts a former ſtate

ſtate of fuſion of all minerals; let us then examine the latter part, in which I conceive the Doctor has egregiouſly deceived himſelf. According to his theory then the minerals of this earth were melted by the combuſtible vegetables that grew on a former earth for an indefinite length of time; this earth, however, the Doctor ſuppoſes to have been inhabited like the preſent *; moſt of theſe vegetables, therefore, muſt have been conſumed, as at preſent, by theſe ancient inhabitants, and many muſt have decayed. There remained only ſuch quantities of them carried into the ſea as are at preſent ſo conveyed, and when there, they muſt have rotted or been decompoſed as at preſent; and only ſuch as eſcaped decompoſition could ſerve as fuel for his mineral heat; but in an indefinite ſpace of time muſt not all of them have been decompoſed? How then, even in that unlimited duration which the Doctor gratuitouſly claims, could ſuch a collection of combuſtibles be accumulated, and in ſuch a perfect ſtate as would be neceſſary to fur-

* Page 177, 188. 273.

niſh a degree of heat ſufficient to melt, I do not ſay a mountain, but even a fragment of quartz? Where has he diſcovered or read that ſuch unmixed ſtrata as would be required in his hypotheſis, were ever found at the mouths of rivers, where ſome part at leaſt of thoſe combuſtibles muſt have been depoſited in the ſame manner as he ſuppoſes them to have been at the bottom of the ſea? How detain the hydrogen? Where find the oxygen, equally neceſſary? On whatever ſide we view this hypotheſis, nothing but improbability or impoſſibility offers itſelf to our view.

The Doctor tells us, p. 143, that "We "muſt not eſtimate the proportion of ma- "terials anciently employed in fuſing mi- "nerals by that which is actually found in "this earth, this he allows is deficient, and "is only the ſuperfluity of that which was "employed." If we are not to judge of the paſt by analogy with the preſent world, I own I am at a loſs how to judge; the Doctor gives no eſtimate, but finds it much more convenient to leave the whole involved in obſcurity.

P. 247, &c. He reproaches me with miſcon-

miſconceiving, or miſrepreſenting, a paſſage of Mr. Dolomieu, wherein I make him ſay, that ſubterraneous heat is not even equal to that of our common furnaces, and the Doctor affirms, that if I had quoted the text inſtead of giving my own interpretation, I could not have offered a ſtronger confirmation of his theory. I ſhall, therefore, now give him the words of Mr. Dolomieu, which were taken, not from the Journal de Phyſique, Mai, 1792, as the Doctor believed, but from the Preface to his Account of the Pontian Iſlands, p. 8; the words are, "Le feu des Volcans *n'a* "*point d'intenſité* il ne peut pas vitrifier les "ſubſtances les plus fuſibles tels que les "ſhorls, il produit la fluidité par une eſ- "pece de diſſolution par une ſimple dila- "tion qui permet aux parties de gliſſer les "unes ſur les autres." It is this laſt opinion it ſeems the Doctor regretted, I had not quoted, as it alludes to a myſterious kind of fuſion, which Mr. Dolomieu then admitted, and which the Doctor conſequently thought favourable to his own, as it participated of the ſame incomprehenſibility. But this great geologiſt has ſince cleared

cleared up this point and perſiſts in denying the great heat of volcanos. I ſhall now quote his own words. Journ. de Phyſ. for 1794. p. 118. Le feu des volcans n'a pas une grande intenſité il ne produit pas une chaleur proportioné à ce qu'on preſumeroit de ſon grand volume, on approche d'un courant de laves ſans eprouver cette ardeur vive & cuiſante que l'on reſſent près des verres & des metaux en fuſion on peut monter deſſus pendant qu'il coule, &c. and p. 121. preſque tous les phenomenes acceſſoires favoriſent mon opinion ſur la fluidité des laves qui ne ſeroit alors *qu'une ſimple ſolution par le ſouffre* & qui n'exigeroit qu'une chaleur *peu ſuperieure à celle neceſſaire pour tenir en fuſion le ſouffre pur.* The inſolent tone the Doctor aſſumes in the ſucceeding pages would call for the ſevereſt reprehenſion did it not ſtill more properly meet it in the ſentiments it muſt naturally excite in the minds of every unprejudiced philoſophic reader.

P. 253—257. He examines my anſwer to the argument he deduced from the apparent fuſion of the native regulus of manganeſe, and concludes by remarking that " my
" obſervation

" obſervation on this occaſion, looks as if I " were willing to deſtroy by inſinuation the " force of an argument that proves the the- " ory of mineral fuſion, and wiſh to render " doubtful, by a ſpecies of ſophiſtry, what, in " fair reaſoning, I cannot deny." To this compliment I ſhall make no reply, but barely ſtate Mr. La Perouſe's teſtimony and the concluſion I drew from it. Mr. La Perouſe relates that he found the native regulus among the iron mines of Sem; that it exiſts in ſeparate lumps like the artificial, but much larger, and that its figure *exceedingly reſembles that of the artificial,* " and that " this exact reſemblance ought, it ſhould " ſeem, to induce us to think it was pro- " duced by fire. That it is very pure, " and contains no part attractable by the " magnet." Hence it is plain Mr. La Perouſe is inclined to think it was produced by fuſion; but I did not think myſelf obliged to adopt this opinion; Mr. La Perouſe's propenſity to believe it a product of fire, was grounded on the great reſemblance of this native regulus to the artificial; this, however, did not convince me, as by the ſame mode of reaſoning, moſt other native

native metals might alſo be aſcribed to fuſion, a notion entertained by none but Doctor Hutton. Local circumſtances muſt concur in ſuggeſting ſuch an opinion; now I knew that here local circumſtances contradicted it; for the manganeſe was found in the mountain of Rancié, a mountain of primitive limeſtone, and among iron ores; moſt of which are hematitic, as Mr. La Perouſe informs us in his Traité ſur les Mines de Fer, p. 8, and 53. neither of which were ever ſuſpected to be of igneous origin except by Doctor Hutton. Hence I concluded the manganeſe could not be deemed to originate from fuſion; the circumſtance of its being found in lumps ſeemed to me a confirmation of this reaſoning, as it is well known that reguli ariſing from fuſion are always diſcrete and ſeparate when the melting heat is not ſufficient, and a defect of this ſort could ſcarce be found in ſuch a heat as Doctor Hutton would have us to adopt, as according to him it could melt quartz, and muſt have melted the primitive limeſtone of which this mountain is formed, therefore the ſize of the maſſes produced makes part of the evidence that this regulus is not a product

of

of fusion, whatever the Doctor may allege to the contrary; but the Doctor answers "that with regard to the nature of the fire "by which the fusion had been produced, "I am much mistaken if I imagine that "the reduction of the reguline or metallic "manganese depends on the intensity of "the heat; it depends on the circum-"stances proper for the separation of "the oxygenating principle of the calx." I suppose he means the reduction of the calx of manganese, for the reguline or metallic manganese being already reduced, requires no reduction; then in opposition to the Doctor I do say that the reduction of the calx or separation of the oxygenating principle *by fusion* so as to obtain a pure regulus does require a very intense heat, and that no known circumstance supersedes the necessity of such heat. This I aver, not only on my own experience, but on that of every chymist in Europe. The words of Bergman are *intensissimo qui in laboratorio parari possit igni exponitur.* 2 Berg. 203. he tells us that he himself could obtain only minute discrete reguli; but that Gahn *by applying a most intense heat* obtained a larger. p. 202. See also

Klaproth

Klaproth 1 Chy. Ann. 1789, p. 11. and 3 Gren. § 3409. therefore the ſize depends on the heat applied. The Doctor cannot ſurely ſuppoſe me a ſtranger to the neceſſity of ſeparating the oxygen. I was the firſt that publiſhed the proceſs for the reduction of manganeſe, in Engliſh in the year 1784. Hence the fallacy of his train of reaſoning and the injuſtice of his reproaches are evident.

P. 258. He reproaches me with not comprehending how coal, an infuſible ſubſtance, could be ſpread into ſtrata by any degree of heat, after he had given three 4to pages endeavouring to explain how all the different degrees of infuſibility were produced; a ſufficient ſpecimen, he ſays, of my underſtanding, at leaſt of his theory. He need not, however, confine himſelf to this ſpecimen, for many other parts of his theory are to me equally incomprehenſible, as I have already often noticed; his explanation of this point in particular, I do not yet comprehend, for all coal appears to me infuſible, except he means by fuſion the intumeſcence that takes place in certain ſpecies of coal from the liquefaction of the bitumen contained in them which makes them cake

but

but not flow, but this partial liquefaction of the bitumen is not a real fusion of the whole compound, but rather similar to that which takes place in borax and some other salts, from water* previous to their real fusion, and has never been confounded with it by any correct writer. Now no one that has ever seen coal thus partially liquefied can comprehend how, if the liquefied part were spread ever so widely, it could form strata of a texture and appearance, so totally different as those we behold in coal mines, nor consequently how the one can be identified with the other.

P. 259. His remarks on my observation on the crystallized *trona*, continued through five whole pages, are too perplexed and tedious to be here repeated. The upshot of my argument was simply this, supposing that *trona* is deprived of its water of crystallization, and yet found crystallized (a circumstance which I had then no opportunity of examining), it would only prove that in this particular solitary instance, that

* And to this Wallerius very justly compares it. 2 Syst. Miner. p. 100.

alkali

alkali was cryſtallized by fuſion, but yet that was not the general mode in which cryſtallized mineral alkali was produced, for that immenſe quantities of it were elſewhere found cryſtallized, but all furniſhed with the water of cryſtallization, and therefore from that ſolitary inſtance which might accidentally be produced, no general inference could be drawn; ſince that time I have on examination found that in the inſtance mentioned by the Doctor, the trona *was not deprived of its water of cryſtallization*, but only contained much leſs of it than uſual. The Doctor, p. 163, cenſures me for not informing him whether thoſe maſſes of mineral alkali which I ſaid to be cryſtallized, retaining their water of cryſtallization were found in what may be properly termed their mineral ſtate, or whether they were transformed from their mineral ſtate by the influence of the atmoſphere. In anſwer to which, I ſhall tell him, that if he means by a mineral ſtate a ſubterraneous ſtate, neither they nor the trona were ſo found; but if he means a ſpontaneous production of nature in the mineral kingdom, they are found ſo circumſtanced, not produced from any ſolid maſs but from a ſtate of aqueous ſolution. The mineral alkali found on the ſea coaſt in India, mentioned in the 6th vol. of the Society of Arts, of London, reſembles trona in ſome reſpects, for Mr. Keir, who examined it, tells us that it is in an intermediate ſtate, between that of cryſtals, which hold a large portion of water, and that of alkali dried as much as it can be. Ibid 141. it is true that this parcel was marked *refined* to diſtinguiſh it from another ſort much more impure, but it does not appear to have undergone any operation of art, in the account given of it, p. 265, &c. The Doctor ſuppoſes the granite which I ſaid was formed in the Mole, conſtructed in the Oder, was nothing more than ſand compacted by mud, and regrets I had not been more particular in my deſcription of it. To ſatisfy him then, I ſhall farther add, that it was ſo compact that it could ſcarce be ſeparated from the real granite to which it was contiguous by a blow, and could not be diſtinguiſhed in colour or coheſion from natural granite, even by the

moſt

moſt experienced mineralogiſt; ſee Prince Gallitzin's 1ſt Letter to Crell, p. 30, or his Treatiſe on Minerals, p. 23. Other inſtances of regenerated granite may be ſeen in my 6th Eſſay, article Granite.

NOTES.

Native Gold.

Page 21. Charpentier ſhews that calcareous earth does not proceed from ſhells, and is often contemporaneous with gneiſs; ſee his deſcription of Saxony, p. 399. 402. 403.

P. 25. Siberia, and probably all other primeval tracts, whether plain or mountainous, were originally much higher than at preſent, having been lowered by diſintegration, to which primeval rocks are moſt ſubject, particularly the higheſt.

P. 46. Some *ſecondary* mountains appear to have been formed by fluviatile inundations, and diſintegration, as the carboniferous, &c.

P. 79. The earthquake that was felt in Canada in 1663, overwhelmed a chain of mountains more than 300 miles long. Clavigero's Hiſtory of Mexico, p. 221.

P. 403. In Mexico, native gold and other metals are chiefly found in ſecondary mountains or hills. Helm, 300. Theſe are undoubtedly thoſe which I call DERIVATIVE, having ariſen from the accumulation of the diſintegrated particles of primeval mountains, as they lie at the foot of, and follow the courſe of, the primeval mountains.

Few parts of Spaniſh America contain maſſes of ſea ſhells. 2 Clavigero's Hiſtory of Mexico, Engliſh edition, p. 249. Theſe ſecondary mountains were not therefore formed under the ſea, but, like the carboniferous, aroſe from the diſintegration of the primeval. The ſame may be ſaid of many of the metalliferous mountains of Siberia and Cornwall, which conſiſt moſtly of ſecondary granite.

8 P. 414.

P. 414. But in Mexico, ſulphurated ſilver ores occur in mountains of granite, gneiſs, and argillite. *Helm.*

P. 417. In Guancavelica in Mexico, a vein of cinnabar, 80 yards thick, is accompanied with galena, manganeſe, and arſenic. *Helm.*

P. 420. In Iouricocha near Paſcho, a belly of porous *brown ironſtone* is found, half a mile long and 15 fathom in thickneſs, containing native ſilver thinly diſperſed through it. But in the *midſt* of it there runs a vein of white argil in which the ſilver abounds. *Helm.*

Is it not evident, that the ſilver originally diſperſed in this porous maſs was conveyed by water into this argil?

P. 421. At Maijos, the ſparry iron ore is found in ſecondary argillitic mountains, accompanied with gold, copper pyrites, and galena.

P. 428. Near Cordova in Mexico, ſome veins of copper ore are found in mountains of red and grey granite. *Helm.*

P. 429. In primeval blue argillite, of which the Cordelieres principally conſiſt, the ſame ores occur, together with thoſe of gold, ſilver, and galena, and the ſparry iron ore, in veins. The famous argentiferous coniform mountain of Potoſi, which is 28 miles in circumference, conſiſts of yellow hard argillite. At *La Paz*, the higheſt point of the Cordelieres, there is an auriferous conglomeration of yellow clay and rounded flints, in a fragment of which, that had lately fallen down, lumps of gold, weighing from two to twenty pounds, were found, and ſome of an ounce weight are ſtill found. *Helm.*

P. 431. In conglomerations of marl, gypſum, limeſtone, and fragments of porphyry, native gold, and ſilver ores abound in the ſtratified mountains of Cuſco. Alſo native

native ſilver, and compact ores of copper and lead. *Helm.*

P. 431. Behind Guancavelica, the argillite graduates into calcareous ſandſtone, as does this into ſimple limeſtone; all equally rich in gold, ſilver, and mercury. *Helm.*

P. 454. Spallanzani has alſo attempted to prove the heat of volcanoes to be very intenſe, but he is refuted by Dolomieu, ſee *Magazin Encyclop.* An. 2d. Vol. I. p. 226.

P. 468. Mr. Hatchett, in a paper lately read before the Royal Society, has ſhewn from the experiments of Mr. Wiſeman, that martial pyrites and ſulphurated ſilver ores, are even now formed in the moiſt way. Phil. Tranſ. 1799. Nay, ſilver that has lain long in the ſea has been found ſulphurated and muriated, though ſulphur can be detected in ſea water by no teſt, which proves the truth of Mr. De Luc's aſſertion, that the ſea may contain ſubſtances as yet unknown. Mr. Gardener informs me, he has found a coating of gold coloured martial pyrites on the ſhells of a ſort of ſhell fiſh called clamp fiſh, in a creek in Eaſt Florida.

P. 492. It may at leaſt be doubted whether the manganeſe in queſtion be in its perfect metallic ſtate, as Chaptal in his Chapter on Ores, ſince publiſhed, has omitted it: But if it be found to be ſo, this proves nothing in favour of Dr. Hutton's Theory, as we may ſuppoſe all metals to have been originally formed in their perfect ſtate.

THE END.

History of Geology

An Arno Press Collection

Association of American Geologists and Naturalists. **Reports of the First, Second, and Third Meetings of the Association of American Geologists and Naturalists, at Philadelphia, in 1840 and 1841, and at Boston in 1842.** 1843

Bakewell, Robert. **An Introduction to Geology.** 1833

Buckland, William. **Reliquiae Diluvianae:** Or, Observations on the Organic Remains Contained in Caves, Fissures, and Diluvial Gravel. 1823

Clarke, John M[ason]. **James Hall of Albany:** Geologist and Palaeontologist, 1811-1898. 1923

Cleaveland, Parker. **An Elementary Treatise on Mineralogy and Geology.** 1816

Clinton, DeWitt. **An Introductory Discourse:** Delivered Before the Literary and Philosophical Society of New-York on the Fourth of May, 1814. 1815

Conybeare, W. D. and William Phillips. **Outlines of the Geology of England and Wales.** 1822

Cuvier, [Georges]. **Essay on the Theory of the Earth.** Translated by Robert Kerr. 1817

Davison, Charles. **The Founders of Seismology.** 1927

Gilbert, G[rove] K[arl]. **Report on the Geology of the Henry Mountains.** 1877

Greenough, G[eorge] B[ellas]. **A Critical Examination of the First Principles of Geology.** 1819

Hooke, Robert. **Lectures and Discourses of Earthquakes and Subterraneous Eruptions.** 1705

Kirwan, Richard. **Geological Essays.** 1799

Lambrecht, K. and W. and A. Quenstedt. **Palaeontologi:** Catalogus Bio-Bibliographicus. 1938

Lyell, Charles. **Charles Lyell on North American Geology.** Edited by Hubert C. Skinner. 1977

Lyell, Charles. **Travels in North America in the Years 1841-2.** Two vols. in one. 1845

Marcou, Jules. **Jules Marcou on the Taconic System in North America.** Edited by Hubert C. Skinner. 1977

Mariotte, [Edmé]. **The Motion of Water and Other Fluids.** Translated by J. T. Desaguliers. 1718

Merrill, George P., editor. **Contributions to a History of American State Geological and Natural History Surveys.** 1920

Miller, Hugh. **The Old Red Sandstone.** 1857

Moore, N[athaniel] F. **Ancient Mineralogy.** 1834

[Murray, John]. **A Comparative View of the Huttonian and Neptunian Systems of Geology.** 1802

Parkinson, James. **Organic Remains of a Former World.** Three vols. 1833

Phillips, John. **Memoirs of William Smith, LL.D.** 1844

Phillips, William. **An Outline of Mineralogy and Geology.** 1816

Ray, John. **Three Physico-Theological Discourses.** 1713

Scrope, G[eorge] Poulett. **The Geology and Extinct Volcanos of Central France.** 1858

Sherley, Thomas. **A Philosophical Essay.** 1672

Thomassy, [Marie Joseph] R[aymond]. **Géologie pratique de la Louisiane.** 1860

Warren, Erasmus. **Geologia:** Or a Discourse Concerning the Earth Before the Deluge. 1690

Webster, John. **Metallographia:** Or, an History of Metals. 1671

Whiston, William. **A New Theory of the Earth.** 1696

White, George W. **Essays on History of Geology.** 1977

Whitehurst, John. **An Inquiry into the Original State and Formation of the Earth.** 1786

Woodward, Horace B. **History of Geology.** 1911

Woodward, Horace B. **The History of the Geological Society of London.** 1907

Woodward, John. **An Essay Toward a Natural History of the Earth.** 1695